Managing the Environmental Crisis in Ghana

Managing the Environmental Crisis in Ghana: The Role of African Traditional Religion and Culture with special reference to the Berekum Traditional Area

By

Samuel Awuah-Nyamekye

CAMBRIDGE
SCHOLARS
PUBLISHING

Managing the Environmental Crisis in Ghana:
The Role of African Traditional Religion and Culture
with special reference to the Berekum Traditional Area,
by Samuel Awuah-Nyamekye

This book first published 2014

Cambridge Scholars Publishing

12 Back Chapman Street, Newcastle upon Tyne, NE6 2XX, UK

British Library Cataloguing in Publication Data
A catalogue record for this book is available from the British Library

ISBN (10): 1-4438-6536-2, ISBN (13): 978-1-4438-6536-4

To

My wife, Joyce, all my children and my beloved mother,
Maame Mercy

CONTENTS

LIST OF TABLES

LIST OF FIGURES

ACKNOWLEDGEMENTS

I wish to express my profound gratitude to the almighty God for the strength He gave me throughout this project. I also acknowledge the moral support I received from my wife, Joyce Effah and all my children.

This book is a revision of my PhD thesis, and as such, I owe a debt of gratitude to my supervisors, Drs. Emma Tomalin and Kevin Ward from the Department of Theology and Religious Studies at the University of Leeds for their painstaking supervision. Dr. Ward, in particular, needs a special mention for spending his energy, time and resources to travel to Ghana to see how I was getting on with my data collection. I am as well indebted to my late cousin, Afari Yeboah (a.k.a. Kwasi Yeboah), who was my able Research Assistant, but who unfortunately did not live to see the end of the project he happily associated himself with when the request was made to him. I am equally indebted to Nana Kwadwo Buadu of Berekum, who readily took over from where Afari Yeboah had left off and tenaciously provided me with the support I needed to see the project to its logical conclusion. Kwadwo, I say, *Ayikoo*! I am also grateful to Dr. Modestus Fosu, who for squeezing some time out of his tight schedule to read the drafts of the various chapters. I am also very grateful to Dr. Andy Cochrane for his excellent proofreading. I am also indebted to Professor Munyaradzi Mawere for critically reviewing the manuscript.

I wish to express my profound thanks to all my key informants–especially those who asked me to use their photos–whose names I cite for their contributions within the book.

A special mention also is given to Professors R. N. Osei and C. R. Gaba, formerly of the University of Cape Coast, for being my academic mentors. Additionally, I acknowledge the moral support I received from Professors B. A. Ntreh and E. B. Anum and Drs. Y. S. Agyemang and P. Appiah-Sekyere and Kofi Appiah all of the Department of Religion and Human Values at the University of Cape Coast

Finally, my thanks go to all those who in one way or the other helped me during the course of collecting data for the book but whom I have not mentioned here.

FOREWORD

BRON TAYLOR

What role if any do religious perceptions and practices play in human relations with the rest of the living world? If religions are or can be a significant variable, how does their influence change over time–as for example, when different people with different religions encounter one another? And how are religious people and their environment-related behaviors changed–if they are– when they encounter those who are deeply informed by evolutionary and ecological understandings, including scientists and activists who promote the conservation of Earth's biological diversity? To what extent, if any, do environmental changes, including those precipitated by human activities, and alarm about such changes, change religions? Does environmental change make religious (and other) people more or less likely to promote behavioral changes that protect the diversity and resilience of the environmental systems upon which all life depends? If human beings are biocultural creatures (which is certainly my view), who are shaped by and shape their habitats in a long, ongoing, process of reciprocal production, then the above questions are critically important, at least if, and when, religion plays a significant role.

Since the mid-19th century certain perceptions about the role of religion in environmental behavior have become widespread. The most common view has been that the Abrahamic religions, and especially Christianity because of its greater global influence than Judaism and Islam in the economic development of the world, harbors anthropocentric attitudes, which in turn yields environmentally destructive behaviors. Often accompanying this view is a perception that religions originating in Asia are more environmentally friendly, especially Buddhism and Daoism. Even more common is a belief that the Earth's surviving indigenous traditions know better than the world's expansionist, agricultural civilizations, how to live harmoniously in their environments. In recent decades some scholars have contended that, whatever their environment-related flaws might have been, the world's religions are becoming more environmentally friendly.

Unfortunately, many of the above perceptions and claims are based on thin evidence, or an unduly charitable approach to the traditions discussed, which is sometimes rooted in an apologetic motivation on the part of the writer to case their own favorite tradition in a positive light. Understandable hopes can also lead observers to downplay, ignore, or fail to notice factors in a tradition that hinder environmental concern and action. Fortunately, hard-headed, on the ground, and empirically oriented research that seeks to uncover the role of the religion variable in environment-related behavior has been increasing. Fortunately, Dr. Samuel Awuah-Nyamekye has taken this approach in his fascinating, ethnographic study of the role of religious beliefs and perspectives in both environmental decline and conservation in Ghana.

While focusing especially traditional religious and cultural beliefs of the indigenous Berekum people (one of the Bono people within the larger Akan ethnic group) in Ghana, Awuah-Nyamekye provides a nuanced, judicious, and complicated analysis. He provides compelling evidence that this indigenous tradition has had a number of important cultural elements that enjoin environmental conservation (even if the conservation impulse is not 'environmental' in a contemporary sense of self-consciously trying to ensure the continuance of all species and ecosystem types). He also contends that these practices have been promoted and enforced by religious authorities and ideas, while acknowledging there are dimensions to these traditions that work contrariwise, or are focused on protecting some but not all species and ecosystem types. He looks forthrightly at the way colonial economic and religious forces have eroded both traditional African religions and lifeways with negative impacts on the regions cultural and biological diversity. But he does not do this in a social science bubble: to his credit he does not shy away from acknowledging the negative ecological and social consequences of dramatically increasing human populations in the regions (and Africa more generally), or the role that political corruption plays in negative ecological and social trends. In short, he does not look at in a simplistic or unidirectional way any of the variables to which he attends. He importantly notes, for example, that the native traditions infuse and at least modestly influence the colonial ones, even while in general retreating in the face of the relative power of their social carriers; and he notes that western scientific and ecological ideas, and conservationist ethics, are exercising their own influences, and that moreover, through a growing appreciation for what is today called traditional ecological knowledge, western scientists are increasingly open to learning from indigenous knowledge and spiritual systems.

Put simply, Awuah-Nyamekye is a keen observer of the processes of hybridity that results when people with differing histories, epistemologies, and spiritualities, encounter one another, especially when they take seriously the beliefs and perceptions of others. I expect it would be fair to say that Awuah-Nyamekye can do this in part because he is himself a byproduct of just such hybridity, as a person from the region he is studying who is intimately acquainted with the peoples and practices there (with unusual access to a wide range of interlocutors), but who earned a Ph.D. at an elite Western university and has been engaged with critical, religion and nature scholars.

As one whose own work is also ethnographic and historical, I know first-hand what a privilege it is to interview people in depth about their perceptions and practices, about what they care about most and why. Reading this book gave me a similar feeling: what a privilege it is to hear the voices of Awuah-Nyamekye and his interlocutors from his homeland. If there is a pathway toward a sustainable future, a way past our fraught histories that have been all too often characterized by the exploitation of one group of people by another, or one part of nature by some group of people, it will involve genuine, honest, encounter, debate, and salutary, mutual, change. It will demand that people with differing epistemological premises take seriously what others claim to know. I will result in new hybrids between the best what people have learned in different times and places.

Despite how challenging the current situation is in Ghana, in Africa, and really, everywhere on Earth, the encounter and hybridity that is unfolding, and which is a part of what Awuah-Nyamekye has documented, is a good sign. I think *Managing the Environmental Crisis in Ghana: The Role of African Traditional Religion and Culture with special reference to the Berekum Traditional Area* is an important contribution to this ferment. It should not to be taken as the last word by any means but rather, carefully considered as a part of an authentic, global, unfolding, cross-cultural, and high-stakes encounter among very different peoples, which, at its best, aims at learning how we might live and let live with our own, and other kind, on the only place in space where we know life exists.

~ Bron Taylor, author of *Dark Green Religion: Nature Religion and the Planetary Future*, and editor of the *Encyclopedia of Religion and Nature,* and the *Journal for the Study of Religion, Nature and Culture. www.brontaylor.com*

INTRODUCTION

Day-in-day-out we hear of concerns being raised across the world about the rate at which our environment is being destroyed and, thus, posing a serious threat to the very existence of different forms of life on earth. As policy-makers, scientists, economists and environmentalists keep on debating the causes and remedies of this crisis, the destruction of the environment continues unabated. This book examines the connection between traditional religio-cultural beliefs and practices as well as environmental problems in Ghana. The book adopts Berekum Traditional Area[1] as a case study. Its primary aim is to assess the nature and the level of the environmental crisis in Ghana, and to explore the possibility of combing indigenous and Western scientific methods to address the current environmental problems in Ghana. Extensive literature (Mbiti, 1991; Menzie 2006; Ntiamoah-Baidu 1991;Nwosu, 2010, Sarfo-Mensah, 2001) exist on the subject matter—'religion and environment/ecology'—which mention that indigenous religions are environmentally friendly or assist indigenous peoples to live in harmony with their environment. However, how these religions assist the adherents to adopt friendly attitude towards nature and how ecological knowledge is generated and transmitted among indigenous peoples are either not examined or are tangentially examined in the existing literature. I find this as a gap. This book aims at filling this gap for, it examines in details how Indigenous African Religion assists its adherents to adopt a friendly attitude towards the environment as well as how traditional ecological knowledge is generated and transmitted among the people of the selected area. Herein lies the uniqueness of this book. This book will be useful to the academics and students in the field of: Religious Studies, Religion and environment/ecology, environmental ethics, indigenous knowledge and those in environmental humanities in general, African studies and policy makers.

---

[1] In Ghana the term 'traditional area' is used to delineate an area where the members of the community are generally homogeneous, or share the same culture, and are under the leadership of an *Ɔmanhene* (Paramount Chief), see clarification of terms section (p.29) for more details.

Background of the book

This book explores the connection between traditional religio-cultural practices and the ecological problems of Ghana. In my previous research (Awuah-Nyamekye, 2009a, 2009b) with some communities among the traditional Akan people[2] of Ghana, I observed that many of these people`s religio-cultural practices have ecological underpinnings. This contrasts with the findings of other scholars such as Alley (1998, 2000, 2002), who argue that in spite of the 'cultic' attention paid to River Ganges as the goddess *Gange Ma*, it is nonetheless dangerously polluted by the Indian population. Tomalin (2002) argues that 'whilst many religio-cultural traditions do have a strong connection with the natural world, elements of the natural world may be considered as sacred without any explicit consciousness about the relevance of this to an environmental crisis' (Tomalin, 2002: 15). For Taringa, the influence of indigenous people's worldviews on human-nature relationships 'are primarily relationships with spirits and not necessarily ecological relationships with nature' (Taringa, 2006: 196). Valid as the above views are, I argue that they may not fully apply to all indigenous societies as I understand societies each society to be unique and heterogeneous.

Based on my field observations and findings, I argue that some of the Ghanaian people's religio-cultural practices can be harnessed to deal with their environmental problems. I must, however, point out that this book is not prescribing a complete retreat into pre-colonial Ghanaian attitudes and practices as this will be anachronistic and also as there is historical evidence to show that an attempt to return to the pre-colonial African heritage that manifested itself in the philosophy of *Negritude*[3] did not appeal to the masses in the 1960s, when even the atmosphere was highly charged against colonial domination (see Dikirr, 2005). *Negritude* did not

---

[2] Those Akan people of Ghana who remain more or less attached to the customs and traditions of their forebears despite the influence of modernity (see the clarification of terms section for more details).

[3] The major motivating force for the Negritude Movement was, according to Mafeje, a protest against 'the disillusionment and resentment of the dehumanizing oppression of colonial domination and suppression of the black people' (see Mafeje 1992). Negritude was championed by African leaders and scholars such as Leopold Senghor, Cheikh Anta Diop, Kwame Nkrumah among others. 'According to Ali Mazrui [2003], the concept of negritude, the celebration of African identity and uniqueness, was invented in Paris by the Martinique poet and philosopher, Aime Cesaire. However, its most famous proponent in Africa according to Mazrui was the founder-president of independent Senegal, Leopold Senghor' (cited in Dirkirr, 2005:118).

appeal much to the masses due perhaps to the fact that its proponents could not explain in clear terms to the people what it was meant for, or perhaps the people thought it was not the right response to their needs at the time. In contemporary Ghana, the combination of Western and Christian influence, together with the effects of globalisation (particularly global capitalism) make such a transition even more untenable. I am also conscious of Dikirr's (2005: 45) argument that 'in today's Africa, a discourse that is wholly predicated on the people's past heritage, especially their alleged spiritual and closeness to the land, will be of little value'. I, however, do not wholly agree with Dikirr though as it will be shown in the ensuing discussions. Rather, the main objective of this book is to bring to the fore what the indigenous ecological practices of Ghanaian people involve, and to examine the extent to which these indigenous ecological methods can complement or blend with Western scientific means of conserving the environment for the benefit of the people in the study area in particular, as well as in Ghana more generally.

Though this book has its main focus on African Traditional Religion, the inroads (influences) of other religions currently operating in Berekum area–most notably Christianity–cannot be ignored. Therefore, the extent to which other religions have affected the indigenous Berekum people`s ecological practices are also explored. Moreover, as there is evidence of hybridity–the fusing of indigenous and other religious beliefs and practices–in the study area and the extent to which hybridity has impacted on indigenous ecological practices with religious underpinnings is also discussed.

The existence of such hybridity seems to corroborate ter Haar and Ellis's (2004) contention that 'religious worldviews do not necessarily diminish with formal education' (2004: 51), and that 'plural religious allegiance is common, with individuals frequenting several religious communities at once or practising rituals regarded by the West as belonging to different or competing systems of belief, such as Christianity and Islam, or Christianity and Traditional Religion' (ibid.: 5). And according to Aylward Shorter (1975: 7), 'the African Christian repudiates remarkably little of his former non-Christian outlook' (cited in Gathogo, 2007: 249). From my interviews, I observed that the most obvious motivation for the people in this study area to blend these worldviews in their practices is what I shall refer to as the 'braces and belt approach to issues'. That is, they do not want to risk adopting the wrong position, and so they tackle the relevant issues by adopting both positions at the same time. Then, if one source fails them, the other is still there as a backup. An important effect of this hybridity is that it may serve as a platform or a starting point for the integration of modern or Western scientific and indigenous methods for addressing contemporary ecological problems. The point here is that since many Berekum Christians accommodate some aspects of the indigenous religion, it is likely that they will not oppose the integration of some of the indigenous ecological practices, which have indigenous religious underpinnings into the modern (scientific) means. Hybridity is discussed in some detail in Chapter Six.

Current environmental situation in Ghana

Those with some knowledge about the state of the environment in the pre-colonial Gold Coast (now Ghana) will appreciate the importance of this book. Ghana is experiencing indiscriminate logging. Miller (2004) reports that the sustainable yield of forest utilisation in Ghana has been exceeded, which has led to environmental degradation (cited in Barre et al. 2008: 27). Ghana experiences annual bushfires, and this is compounded by illegal surface mining, poor farming practices, and the dumping of human and industrial waste into bodies of water. Current records estimate that over 90% of Ghana`s high forest has been logged since the late 1940s (Tamakloe, 2008), and the rate of deforestation is 5% per annum outside of reserves (Tamakloe, 2008). The above, indeed, is one of the key motivations for this book.

The first state of Ghana's Environment Report was launched in 2004, which emphasised that:

> The weather is becoming warmer with rainfall patterns fluctuating over the last 40 years, with gradual decreases in rainfall distribution in all parts of the country. ..climatic variability held consequences for land use patterns especially in agricultural areas as well as the increasing rate of deforestation (Ghana News Agency [GNA], 2004).

Launching the report, Ms Christine Churcher, the then Minister of Environment and Science, said that 'the environmental impact of land degradation included reduced crop yield, reduced quantity of vegetative cover and reduced resilience of land to climatic variability' (GNA, 2004). The 130-page report dealt with issues such as atmospheric quality, land use and land cover, fresh water, forestry, the coastal and marine environment, bio-diversity, energy and human settlement (GNA, 2004). The startling revelation of the report was that 'the original forest cover was about 36% of the country's landmass, reducing to 23 % by 1972, 13.3% in 1990 and 10.2 % 2000' (GNA, 2004).

'It has been assessed that about 30% of the forest areas are destroyed by fire each year and only about 20% of the forest zone is currently covered by forest that has not burnt regularly.' (Agyarko, 2001). Intentional fires, started to reduce the ferocity of late dry season vegetation fires near the forest-savannah boundary have had a negative influence on forest regeneration, and hunters who fail to extinguish campfires are a major source of unintentional fires (see Agyarko, 2001). It is estimated that more than one million cubic metres of exportable timber have been lost to fire over the past decade (The New Legon Observer, 2008: 3; see also Agyarko, 2001). There is the Provisional National Defence Council (P.D.N.C.) Law 229 in place to check bushfires, but this law does not have enough 'bite' to deal with bushfire offenders as many bushfire offenders mostly go unpunished. Moreover, the District Environmental Management Sub-Committees (DEMC)–the main environmental bodies within the various Assemblies[4] in Ghana–is weak, partly due to under-resourcing.

It is thus no surprise that the original forest cover, which was estimated to comprise approximately 36% of the country's land-mass, was rapidly reduced to 10.2% of land cover by 2000–that is to 2.1million hectares of Ghana`s original 8.1million hectares of moist forest (Dorm-Adzobu et al., 1991). Moreover, it has been estimated that environmental degradation's cost to the economy of Ghana is 10% of its annual GDP ($1.2 billion of its $12 billion GDP), while water and air pollution, deforestation and desertification also continue to take their toll (GNA, 2007).

The World Bank's Water and Sanitation Programme (WBSP) reported that Ghana's economy loses $90 million annually (GHC420 million–1.6 % of GDP) through poor sanitation (GNA, 2012). The GNA report further that a survey carried out in 2008 on sanitation in Africa ranked Ghana 14th out of the 15 countries in West Africa and 48th out of the then 52 (now 53) countries on the continent in terms of sanitation, with the worst progress in the development of sanitation (GNA, 2008).

Various governments of Ghana have made several attempts to salvage this situation, but the desired results are yet to be achieved. As I have earlier on noted, my preliminary studies among some traditional Akan people of Ghana identified that these people`s indigenous ecological knowledge, which is underpinned by their religion and culture, could be of help for dealing with some of the environmental problems of today. As I have earlier on noted, my preliminary studies among some traditional

---

[4] Ghana has been zoned into ten administrative regions. Depending on the population size, the regions are also divided into Districts, Municipalities and Metropolitans.

Akan people of Ghana identified that these people`s indigenous ecological knowledge, which is underpinned by their religion and culture, could be of help for dealing with some of the environmental problems of today. This is because they have developed mechanisms that inspire them to live lifestyles that promote conservation of biodiversity. Such mechanisms include reverence for various bodies of water and areas of land (including the protection of sacred groves) and the belief in *sasa* (the belief that some plants and animals have powerful spirits that need to be propitiated and treated with caution to avoid negative consequences), and these mechanisms are mediated by the institution of chieftaincy. This suggests that their worldviews could play a major role in improving their ecological practices.

In Ghana, however, conservation policies often neglect the indigenous people`s ecological knowledge or worldviews. A clear evidence of this is that an important document like *National Biodiversity Strategy* for Ghana was completely silent on the indigenous people's knowledge in this matter (Ministry of Environment and Science, 2002). Not only Ghanaian policy-makers, but also many of the World Bank's interventions on sustainable development for Africa (such as the World Bank's Environment Strategy (WBES) for Africa, and the Environmental Initiative of the New Partnership for African Development) do not factor religion into their schemes. Although the Brundtland Report did recognise the potential for indigenous knowledge to play a role in sustainable development, this recognition was still 'framed within the dominant Western agenda' (McGregor, 2004). The reason for this may be due to the perception that there is no scientific basis for indigenous conservatory methods (Awuah-Nyamekye, 2009a: 30). Or it may be due to what Tomalin (2009: 2) views as being 'grounded in the modernist assumption that religion will disappear once societies "develop" and become more like the West.' Attuquayefio and Fobil (2005) were emphatic that many of the environmental conservation initiatives in developing countries (particularly in Africa) have not been effective because of the tendency to overlook the essential link between traditional and scientific ways of conserving the environment. Also, as Boersema and Reijnders (2009: 4) have argued, when it comes to solving environmental problems–which are ultimately social problems–science and scientists' capabilities become limited. This, according to Boersema and Reijnders, is a matter that relies most importantly on funding, political will and the cooperation of stakeholders.

I have already noted that in view of the environmental problems in Ghana, religion and culture could play a major role in the process of environmental conservation among the Akan. This is because the traditional Akan and, for that matter, traditional Ghanaians, believe that they have a deep responsibility to act as good environmental stewards, accepting personal responsibility for environmental accountability (Beavis, 1994). This also implies responsible management of natural resources for the benefit of present and future generations. Therefore, I argue that Ghana's indigenous religions and cultures are potential resources, which, if well considered, could complement the government's efforts to deal with Ghana's contemporary environmental problems. Moreover, the abandonment and modification of theories of secularisation due to the 'resurgence of religion' and the increased public interest in environmental issues (see Tomalin, 2009: 2; Haynes, 2007: 1) makes this book important. It is against this background that I have decided to use Berekum Traditional Area as a case study to investigate in detailed how indigenous religion and culture could be used to address some of the environmental problems of Ghana.

Summary

In this introductory section, I have outlined the subject matter of the book–religion and the environment. The background or the motivation for the book has been clearly spelt out, and have in addition, presented an overview of the current environmental situation in Ghana to set the basis of the book.

I have also listed some of the indigenous conservation strategies (to be discussed in detailed in chapter 5). I have furthermore pointed out that conservation policies in Ghana often neglect the indigenous people`s ecological knowledge or worldviews.

Besides, I have argued that Ghana's indigenous religions and cultures are potential resources, which, if well considered, could complement the government's efforts to deal with Ghana's contemporary environmental problems.

CHAPTER ONE

HISTORICAL BACKGROUND: MODERN/RELIGIOUS ENVIRONMENTALISM

Introduction

The focus of this chapter is to outline the trajectory of the 'modern' environmental activism and how religion came into the picture-religious environmentalism. I also present a conceptual analysis of the central themes or terms-*African Traditional Religion, environmental crisis and culture*—that feature prominently in the discussion.

The genesis of modern environmental activism

In order to place the discussion in historical context, I will briefly trace the emergence of modern environmental or activism. Scientists, environmentalists and concerned people have become increasingly apprehensive about the 'problems of resource depletion in combination with population growth and world hunger, problems of climate change, the extinction of species, the unsafe storage of nuclear waste, acidification, ozone layer depletion, dehydration, the pollution of air, water and soil and natural degradation' (Postma, 2006:.4). They feel that there is an urgent need for something to be done to address this alarming situation in order to save the world from destruction.

The history of responses to the environmental crisis is complex, but I must point out that environmental activism in the U.S.A–whose origins date back to the nineteenth century–played a major role in it. This activism resulted in the publication of Rachel Carson`s *Silent Spring* in 1962, which sought to draw attention to the damaging impact of pesticides on humans and the environment (Dunlap and Mertig, 1991: 1-3). Moreover, Cahn and Cahn (1990) claim that the Earth Day Conference organised in 1990 by the US Environmental Movement '...united more people concerned about a single cause than any other global event in history' (Cahn and Cahn, 1990: 17, quoted in Dunlap and Mertig, 1991: 1). The publication of Lynn White`s

The Historical Roots of our Ecological Crisis in 1967 – which sought to blame Christianity for the world`s environmental crisis– indeed, drew religion into the environmental debate.

Although White`s thesis has been questioned by others for being overly simplistic (Johnson and Butigan, 1984; Harrison, 1999; Johnson, 2000; Attfield 1983, 2010), it is still relevant to today's debate. As confirmed by Leslie *et al.* (2007), for instance, that White`s 'analysis generated a major controversy that continues and stimulated the growth of the fields of environmental ethics and ecotheology'. Even White himself recognised that religion has a role to play in the solution to the environmental crisis (Leslie, 2007). Indeed, he went even further, to state that 'what people do about their ecology depends on what they think about themselves in relation to things around them. Human ecology is deeply conditioned by beliefs about our nature and destiny–that is, by religion' (Lynn White, 1967: 1204). White's publication is significant, for what started as a debate has now metamorphosed into an important sub-field–religion and the environment–in the academic disciplines of religious studies, theology, ethics, and the sociology of religion, among others.

I argue that all the above forms of environmental activism no doubt influenced the creation of environmentally-based international treaties or conventions such as The Ramsar Wetlands Convention 1971; The World Heritage Convention 1972; The Convention on International Trade in Endangered Species (CITES) 1973; The Convention to Combat Desertification (UNCCD) 1974; The Convention on the Law of the Sea (UNCLOS) 1982; The International Convention on Oil Pollution Preparedness the Response and Co-operation (OPRC) 1990; as well as international conferences, such as the UN Conference on the Human Environment in 1972, the Assisi Conference of the International World Wildlife Fund (WWF) in 1986, and The Brundtland Commission in 1987. These culminated in the famous 'Earth Summit' (UN Conference on Environment and Development) in Rio de Janeiro, Brazil, 1992.

These global activities have resulted in a rise in environmental consciousness, as evidenced in many publications[1] and conferences,[2] as well as a rise in the interest concerning the relationship between religion and the environment. This is seen in organisations such as the Alliance for

---

[1] See for instance: Hays (1987), Cooper and Palmer (1998), McNeely (1999), Dodds (2000), Ellen, Parks and Bicker (2003), Menzies (2006), Burley, Helen and Haslam (2008), Ntiamoa-Baidu (1995, 2008), Tucker and Grim (2000) and Attfield (2010). Religions of the World and Ecology series edited.

[2] Including the World Summit on Sustainable Development, Johannesburg (2002) and the Biodiversity Conference, Nagoya, Japan (2010).

Religion and Conservation (ARC) based in Manchester, UK (founded in 1995); Forum on Religion and Ecology which was announced at the United Nations in 1998 and was embedded in Yale's School of Forestry and Environmental Studies in 2006, now Yale Forum on Religion and Ecology. An important contribution to the debate came from the "Religions of the World and Ecology Conference Series", which took place at the Center for the Study of World Religions (CSWR) at Harvard Divinity School and covered a period of over three years (1996-1998). The papers presented by more than 800 experts-- scholars, religious leaders and environmentalists-- were edited by Mary Evelyn Tucker and John Grim and published in a series under the titled: World Religions and Ecology Series.

Different theories have been propounded to explain the causes of and the possible solutions to the environmental crisis (Attfield, 2010). Park (2007: 5), for instance, contends that 'whilst the environmental problems seem to be physical, their causes and remedies can be found in the "people's attitude, values and expectations"'. This implies that the perceptions or worldviews of a people may influence their behaviour towards environmental issues. Hence, a chapter of this book has been devoted to examining the worldviews of the people being studied in this book.

Park (2007) cites the realisation of the following three ideas as triggers for the contemporary interest in environmental issues: (a) the fact that 'most human activities affect the environment in one way or another, usually for worst', (b) the realisation that 'the environment is our basic life-support system. It provides the air we breathe, the water we drink, the food we eat, the land we live on', and (c) the fact that 'many parts of the environment have been badly damaged by over-use or unwise use [of it]' (Park 2007: 5).

The nexus between religion and the environment

The idea that the connection between religion and environment deserves critical study has been pointed out in the *Religions of the World and Ecology series.* This was a series of conferences hosted at the Centre for the Study of World Religions (CSWR) at Harvard Divinity School. These conferences covered over three years (1996-1998). The papers presented by more than 800 experts-- scholars, religious leaders and environmentalists-- were edited by Mary Evelyn Tucker and John Grim and published in series under the title: World Religions and Ecology Series.

Sequel to the environmental debate and activism was the rise in the interest of the relationship between religion and the environment, which many commentators refer to as the emergence of religious environmentalism and religious environmentalist discourse. Gottlieb (2009) for instance, sees religious environmentalism as a 'diverse, vibrant, global movement" of ideas and activism that "roots the general environmental message in a spiritual framework' (2009:17, 215, 231). The discourse has resulted in a body of literature examining the relationship between religion and the environment that emerged within the context of the West, but is now largely globalised.

Initially, what was happening elsewhere in the world, including Africa, received little attention within this discourse, even though the poorest countries bear the brunt of climate change and its attendant effects. Tomalin (2009: 36) quotes Patz *et al.* (2005) as predicting that:

> Although the world`s poorest countries produce less greenhouse gases than the wealthier nations, it is within these regions that dramatic rise in diseases and malnutrition, as a result of climate change, are likely to have a devastating impact: "the regions with the greatest burden of climate – sensitive diseases are also the regions with the lowest capacity to adapt to new risks (Patz et al. 2005: 315).

The issues of religious environmentalism in the poorest countries of the world–especially those in Africa, south of the Sahara – should have received much more attention from the developed world from the start. As Tomalin (2009: 36) rightly points out, 'in Britain, for instance, environmental concerns are rarely survival issues, whereas in India [and in Africa too], environmental problems (such as the contamination of water supplies in urban centres or effects of rural deforestation on farming practices) are also direct life and death issues'. However, as has been noted, the debate now has a global dimension, and the role of indigenous religions in environmental conservation has been drawn into it. It is possible that scholars–particularly those from Africa, most of whom are Western trained–are responsible for extending the environmental debate to Africa. The story of religious environmentalism will not be complete if the key role played by the Assisi meeting of 1986 is not mentioned.

The Assisi Declarations

The celebration of the 25th anniversary of the World Wide Fund for Nature (WWF) in Assisi, Italy in 1986 also recognised the relationship between religion and the environment. The WWF emphasised the idea that each religious tradition has something positive to offer its adherents in their quest to resolve the environmental problems confronting them. Therefore, under the auspices of its President, HRH Prince Philip, the Duke of Edinburgh, the WWF and (as part of the celebrations) leading members from five major world religions–Buddhism, Christianity, Hinduism, Judaism and Islam–were invited to declare how their various faiths help them to care for nature. This event is today known as the 'Assisi Declarations',[3] and is very important within the field of religious environmentalism, because interest in the field gathered momentum from this time onwards, with the declarations leading to the creation of a network of faith groups working on ecological and developmental issues. By 1995, Baha'i, Daoism, Jainism and Sikhism had also made their declarations on nature conservation, and in the same year a new NGO called the 'Alliance of Religions and Conservation' (ARC), which aimed to support the growth of the network, was established (Wolfensohn, 2003).

One thing that is fundamental to the declarations that were made is that they depict the histories of the world religions as defining the place of humans in nature, including how they should act toward nature. This may be why Leslie et al. (2007) claim that 'whatever religious people consider to be sacred or spiritual is more likely to be revered, protected and conserved' (Leslie et al., 2007; see Hagan, 1999; Nsiah, 2009).

Although indigenous peoples (including Africans) were not at the Assisi meeting, some studies argue that they have been living in harmony with nature since time immemorial (see Mbiti, 1991: 44). The following statement from one of my key informants, which became evident throughout my interviews reflects the traditional Berekum people's understanding of nature:

> In God's creation there are things that can be used (destroyed) and other that cannot be used, so the destruction of such things constitutes a sin. Even those that can be used have the right time for their use. Anything short of this is considered as a sin in our tradition. This is to ensure that we make sustainable use of natural resources (Akumfi Ameyaw, personal communication, 7 November 2011).

Park (2007: 4) says 'since the 1970s, scientists have been writing about what they term the "environmental crisis" and more recently this debate has evolved. This study represents the author`s contribution to this debate

---

[3] For the actual declarations see http://www.arcworld.org/downloads/THE%20ASSISI%20DECLARATIONS.pdf

from the Ghanaian perspective, utilising the Berekum Traditional Area as a case study.

Conceptual analysis of terms

In order to assist readers in developing a clear understanding of the arguments in this book, it is prudent to first clarify some of the terms that will feature prominently in course of the discussions.These terms include: tradition, Berekum Traditional Area, Indigenous, Traditional/ Indigenous Ecological Knowledge, Traditional African Religion, Culture, and Environmental crisis.

Tradition (Traditional)

I use 'tradition' to refer to the long-established patterns of behaviour that have been handed down from generation to generation in Berekum society. These long-established patterns of behaviour are known as *amanmre* in the Akan language. To say 'handed down from generation to generation' does not necessarily mean that this tradition is monolithic, and there is in fact, clear evidence of contemporary manifestations in Berekum people's way of life, resulting from the impact of colonialism and their contact with other groups of people. In fact, there is an element of dynamism in this society, with components of continuity and discontinuity. For instance, human sacrifice is no longer practised in Berekum society, and fewer people are still living in thatch houses as there were in recent history. Also a considerable number of people are now able to read and write and enjoy many modern social amenities, such as electricity, piped water, and modern methods of health care.

Berekum Traditional Area

To refer to the Berekum society as 'traditional' means, to borrow Opoku's (1978) words, that Berekum society 'is undergirded by a fundamentally indigenous value system and that it has its own pattern, with its own historical inheritance from the past' (Opoku, 1978: 9). Therefore, when an area is described as 'traditional' in Ghana, this is used to express the idea that the people of that area share a common heritage that is bequeathed to them by their forebears. In the traditional Ghanaian political system, when an area is described as 'traditional', it means that the place is *autonomous*. That is, it qualifies as an *ɔman* (state), with its own leader, known as a *Ɔmanhene* (a paramount chief).

Indigenous

Although there is some debate over the scope of the term 'indigenous', my use of it here stems from its etymological sense – that is, from its Latin origin – *indgena* (native). Therefore, in this book 'indigenous' refers to those people whose forebears founded settlements where they still remain today. In other words, indigenous people are those that can lay legitimate claim to the land they are occupying today through their ancestry. In this way, one may also refer to indigenous people as 'traditional' people. For this reason, the two terms are used interchangeably in this book (adapted from Awuah-Nyamekye and Sarfo-Mensah, 2011).

Traditional/Indigenous Ecological Knowledge

Traditional or Indigenous Knowledge, particularly in relation to Indigenous people and the environment has become a subject of research in recent time. However, some scholars see the term Traditional or Indigenous, as ambiguous (see Berkes, 2008: 3). This, thus, makes the application of the term (Traditional or Indigenous) to ecological knowledge more problematic. In view of this there is no universally accepted definition of what we refer to now as 'Traditional' or 'Indigenous' Ecological Knowledge. Menzies(2006) cites Bombay (1996) quoting Jameson Brant, a Mohawk as having described Indigenous knowledge as 'A body of information about the interconnected elements of natural environment which traditional Indigenous people have been taught, from generation to generation, respect and give thanks for' (cited in Menzies, 2006:.6). Berkes (2008) and Menzies, (2006) have discussed in some details some of the definitions and attributes of Traditional Knowledge. Menzies for instance, observes that Traditional Ecological Knowledge is a term 'used to describe the knowledge and beliefs that Indigenous people hold of their environments that is handed through the generation (Menzies, (2006: 6). In this book, the terms Traditional or Indigenous Ecological Knowledge refers to the knowledge that local people have built over the years about their environment–particularly in relation to the local flora, fauna and other natural phenomena–and the means designed (with some having religious underpinnings) to interact with these natural elements in a sustainable way.

Traditional African Religion

Many of the earlier writers who briefly visited Africa thought that Africans had no religion. This may be due to the fact that they were expecting to see priests in charge of the worship of the Supreme Being (God), places of worship dedicated specifically to the worship of God (such as temples, cathedrals, or mosques) and some sort of written scriptures (such as the Bible, Quran, or the Bhagavad Gita). As they did not see any of these things, they presumed there is no organised public religion in Africa. But later, writers such as Parrinder, Rattray, who actually had the opportunity to stay longer and study the people, realised that indigenous Africans did have a religion. Rattray for instance, had the opportunity to spend a number of years with the Asantes of Ghana. He made the following remarks about African Traditional Religion in general and in particular the Asantes of Ghana:

> I sometimes like to think, had these people been left to work out their own salvation, perhaps someday an African Messiah would have arisen and swept their pantheons clean of the fetish (suman). West Africa might then have become the cradle of a new creed which acknowledged One Great Spirit, Who, being One, nevertheless manifested Himself in everything around Him and taught men to hear His voice in the flow of His waters and in the sound of His winds in the trees (Rattray, 1959, P. iii).

Although this comment suggests that Rattray perceived African people's religion in polytheistic terms, the point is that he did see that African people do have religion. Parrinder, however, had a better understanding of the African traditional religion. He remarked that:

> It is probably true to say that African religion has been more misunderstood, and has suffered more at the hands of the early writers, than any part of African life. Unhappily old misconceptions linger with us still (Parrinder, 1974, p. 13).

The above observation is appealing and the impression I gathered through my readings prior to my training in the Department of Religion and Human Values at the University of Cape Coast in Ghana was that the spirituality of indigenous African peoples could not be referred to as *religion* in the strictest sense of the word. However, my studies on the spirituality of the African peoples led me to realise that the spirituality of its indigenous peoples can be referred to as a religion, because a close analysis of the belief systems of these peoples indicates that they had most of the elements that most of the Western scholars refer to as religious. One particular book that shaped and deepened my understanding of African religion was *African Traditional Religion* (1974 [1951, 1962]), which was written by Parrinder–a non-African. His book is one of the expositions on African Traditional Religion that paved the way for the recognition and acceptance of African Traditional Religion as a religion in its own right. Rejecting the obnoxious terminologies used by some of the earlier Western writers to denote the spirituality of Africans, Parrinder argued:

> It is evident that fetish is a most ambiguous word [for African religion] and the time has come for all serious writers and speakers to abandon it completely and finally (Parrinder 1974: 16).

However, my research and experience show that even today, there are writers (including some Africans) who continue to see African religion as a false religion and use early European terminology to describe it. However, whatever such authors say about the spirituality of the Africans, there is enough evidence for African spirituality to be justifiably referred to as religion. Therefore, for the purposes of this book, African Traditional Religion will be understood as the religion of the people of the indigenous Berekum people, with all its current manifestations, in contrast to other religions such as Christianity and Islam

Scholars on African religion generally agree on what this religion involves, as the above survey shows, but disagree when it comes to the matter of whether it should be understood in the singular or in the plural. After struggling to debunk inaccurate and pejorative terminologies such as paganism, fetishism, ancestral worship, witchcraft, animism and others that were used by early Western writers (especially anthropologists), African scholars (theologians, anthropologists and sociologists) are now confronted with the difficulty of categorising and naming African religion.

There is a longstanding debate over whether Africans have a single religion or a plurality of religions. In the words of Ejizu (not dated), this may sound rather elementary, but the truth is that the issue of the precise name and nature of the spirituality of the indigenous Africans remains unsettled (Ikenga-Metuh, 1987: 19). Ejizu intimates that this debate has engaged scholars and been the focus of publications and conferences for a considerable period of time. The school of thought led by Idowu (1973: 103) argues that it is correct to talk about 'African Traditional Religion' in the singular form, whilst that led by Mbiti insists that it is deceptive to speak of the spirituality of the African in singular terms. For him, 'we speak of African religions in the plural because there are about one thousand African peoples (tribes), and each has its own religious system'

(Mbiti, 1969: 1-2). Mbiti is supported by scholars such as Ray (1976) and Ikenga-Metuh (1987), who argue that one cannot truly speak of a uniform 'African Traditional Religion' because religion in Africa is tribe-based, with each religion being limited to the ethnic group within which it has evolved. One traditional religion cannot be propagated in another tribal group, according to this view, and moreover, there is no conversion from one traditional religion to another

The pro-single religion group is also of the candid opinion that there is evidence to show that one can genuinely speak of African religion in a homogenous sense, as there are a remarkable number of features of religious beliefs and practices that are found everywhere in sub-Saharan Africa, as well as a unified basic world-view.

According to Westerlund (1991: 16), comparative research conducted by scholars on religion revealed that 'there seems to be a search for transcultural "spiritual unity", and a tendency to emphasize similarity rather than dissimilarities'. That is, those who emphasise the homogeneity or similarities and thus argue for a singular term may have been influenced by the above-mentioned view. Some scholars have even argued that "the influence of nationalism cannot be overlooked" (Westerlund, 1991: 19). The contention is that some post-independence scholars of African descent have tended to present Africa as a unified body including her religious expressions. This might have influenced those who would like to refer to African religion in singular terms.

As the debate rages on, the progenitor of the heterogeneity theory on African Religion, Mbiti became converted and said there is indeed a basis to refer to African religion by a singular term. Interestingly, his underlying argument for his change of mind is that African spirituality cuts across the African sub-region and shares the same worldview. For instance, he observes that, 'in many African languages, the name of God means 'Creator'; even where there is another name; He is called "the creator" as well' (Mbiti, 1991: 35). Thus, he now accepts the position Idowu and his supporters have expounded all along (see Alolo, 2007). And indeed he went on further to title one of his several books on African spirituality as *Introduction to African Religion* (in the singular form). He published this book in 1975 and revised it in 1991. This seems to have given the supporters of the singular school of thought an upper hand. But the reality on the ground points to the contrary.

The view of the pro-singular African scholars on African Religion is expressed in the following quote from Ejizu (not dated):

> The existence of a common world-view as well as similarities in belief-systems, ritual forms, values and institutions across the various regions of

> the continent, provide a sufficient basis for keeping the singular form of the name. There is no reason to single out African Traditional Religion, while accepting as normal a multiplicity of denominations, even rival sects in other religions of humankind, including Hinduism, Buddhism, Judaism, Christianity, and Islam.

It may be reasonable to suppose that the above view may have played a role in Mbiti's conversion. It is interesting to note, however, that despite the repudiation of the heterogeneity theory by its originator, the debate nonetheless continues. Contemporary scholars on African religions feel that in spite of the change in mind of the proponent of the plural form of African religion, there are still good reasons to refer to African religions in the plural form rather than in the singular. I observed one of the clear evidences of this when I sent an article to a journal for review and publication in South Africa in 2009, and one of the comments from the assessors was that there is nothing like an 'African Traditional Religion' but only 'African Traditional Religions'.

The debate, however, continues unabated. Ejizu (not. dated), for instance, argues that contemporary scholarship seems to favour the singular form while Shoko (2012) contends that 'the debate between singular and plural formatives tilts towards the plural' (Shoko, 2012: 54). However, it is common to find many contemporary scholars using both the singular and the plural forms in discourses on African Traditional Religion. At times, one can find one scholar using both terms interchangeably in his/her writing even on the same page. For instance, as late as 2012, Danfulani, a respected scholar on Africa religion uses both terms in his excellent piece titled 'Africa religions in scholarship: A critique', which appears in the first chapter of an edited volume by Adogame et al (2012)—'African traditions in the study of religion in Africa'--Danfulani uses both terms on the very first page of his discussion--p. 19--as well as on pp, 23, 25 and 28. Altogether, he used the singular form more than the plural throughout the discussion. In the same volume, Togarasei (2012) uses the singular form throughout in discussing his paper titled: 'African Traditional Religion in the study of the New Testament'. What this means is that the debate will rage on for a very long time since to some extent, there is basis to use either of them depending on where the emphasis is being placed. In other words, it 'depends on the individual scholar's worldview and interest' as Westerlund (1991: 18) suggests. In this book, however, the singular form is preferred—African Traditional Religion since I identify myself with the pro-singular group.

Culture

Since this study examines the Berekum religio-cultural contribution to nature conservation, it is also important to look at the concept of culture. The word 'culture' was introduced into anthropology as a technical term by Tylor (White and Dillingham, 1972:. 21), who used it in the opening words of his widely read book *Primitive Culture* in 1871. Tylor saw culture as, 'that complex whole, which includes beliefs, art, law, morals, customs and any other capabilities and habits acquired by man as a member of society' (quoted in White and Dillingham, 1972.: 21). Storey (2001) cites William Raymond as having conceptualised the term 'culture' in three different ways. First, it can be viewed as 'a general process of intellectual, spiritual and aesthetic development' (Storey, 2001: 1); secondly, it can be used to refer to 'a particular way of life, whether of a people, a period or a group' (Storey, 2001: 2); and third, it can be used to refer to 'the works and practices of intellectual and especially artistic activity' (Raymond, 1983: 90; quoted in Storey, 2001: 2).

Geertz (1973) thinks that we should see cultures, 'not as organically unified or traditionally continuous but rather as negotiated, present processes' (Geertz, 1988: 273). He also quotes Kluckhohn as describing or defining culture in his book '*Mirror for Man*' in at least eleven different ways: (1) 'the total way of life of a people'; (2) 'the social legacy the individual acquires from his group'; (3) 'a way of thinking, feeling, and believing'; (4) 'an abstraction from behaviour'; (5) 'a theory on the part of the anthropologist about the way in which a group of people in fact behave'; (6) 'a store-house of pooled learning'; (7) 'a set of standardized orientations to recurrent problems'; (8) 'learned behaviour'; (9) 'a mechanism for the normative regulation of behaviour'; (10) 'a set of techniques for adjusting both to the external environment and to other men'; and (11) 'a precipitate of history' (Geertz, 1973: 5). Mugambi (1996: 32), however, sees culture as having six main pillars–politics, economics, ethics, aesthetics, kinship and religion. Mawere (2014) however adds two more to the six components--technology and ideas/abstractions/ imagination/thinking.

Mbiti (1991) characterises this as covering a broad range of things, manifesting itself in a people's art, literature, language, dance, music, drama; the style through which they build their houses and dress; the way they organise socially and politically; their religious ideas, ethics, moral and philosophy; the way their customs, economic life and institutions of people are structured and their values and ethic (Mbiti, 1991: 7).

Dogbe (2009: 101) says that the term 'culture' brings to 'mind many things including ideas, beliefs customs and practices', which cover areas such as 'what life should be and should not be, the people`s judgement, inventions, languages, technology, art and craft, literature, music, ways of farming, building houses, storing and preparing food, weaving clothes and other things'. Mugambi, Mbiti and Dogbe's descriptions of culture are very comprehensive, including almost all its essential elements–both the tangible or material (tools, machines, food, clothing, houses tools for leisure and entertainment) and the intangible (institutions for governing and maintaining law and order, belief systems, values, history, ideologies and concepts). Their descriptions also satisfy the various nuances that William Raymond suggests the term culture covers.

According to Peter Sarpong–a renowned Ghanaian theologian and an authority on Akan culture–'culture is the integrated sum-total of behavioural traits that have been learned, and […] passed on from one generation to another in an uninterrupted succession' (1974: vii). He quickly adds that 'all the same, the transformation of culture cannot be so profound as to leave nothing of it', even though 'culture is dynamic and never static' (1974: vii).

Gathogo (2009: 83) corroborates this when he argues that, 'while cultural meanings may differ historically, it would however be erroneous to suggest that there is a complete break up in cultural meanings, as there is also continuity'. Flolu (2009:. 59) reiterates this, noting that all cultures are subject to growth, change and development. One thing that emerges from the various definitions of culture is that it is an important element in every society. In other words, culture is generic to humanity, as it is that which distinguishes one group of people from another. For this reason, it is argued in this book that any group of people that loses its culture loses it identity. This is one reason why as has been pointed out earlier, it is highly unacceptable for any group of people to suppress another`s culture even if those whose culture is being suppressed will gain materially. This is because culture is an inalienable right which must be jealously protected at any cost. The importance of culture is crucial to peoples all over the world. This explains the reason why people have gone to war in the name of defending their culture against imperialism (e.g. the Meccabbean revolt against Antiochus Epiphanes when Antiochus tried to impose Hellenistic culture on the Jews).

This shows that it is culture that influences or directs members of a society to achieve or realise their goals and aspirations in life.

Dogbe (2009) also makes the important observation that there are some cultural objects that have unique characteristics which render them unsuitable for use in other cultural settings–for instance, some European clothes and houses are designed for cold winter weather, and are hence useless in countries that have tropical climates. The implication of this is that it is totally wrong for any group of people to impose their culturally grounded values and practices on others. This is in line with Tomlinson's criticism of 'personal observation' methods, which tend to be paternalistic and presume that those in the West know the needs of other cultures better than the members of those cultures themselves do (quoted in Gathogo, 2009: 87).

Religion is another important component of every culture, and as such cannot be ignored. Mbiti (1991: 10) argues that religion 'is by far the richest part of the African [people's] heritage'. This is because religion is prevalent in all spheres of human life (Mugambi, 1996). That is, it shapes their cultures, their social life, their politics, and economics (Mbiti (1991:10).

The foregoing discussion can leave little doubt that there is a significant debate over what the term 'culture' refers to and how it evolves over time. It is no wonder that William Raymond describes culture as 'one of the two or three most complicated words in English Language' (Raymond, 1983: 87, quoted in Storey, 2001: 1). Culture, to borrow Storey`s words, may be seen as 'an empty conceptual category, one which can be filled in a wide variety of often conflicting ways, depending on the context of use' (Storey, 2001: 1). This also suggests that the concept of culture is very broad, even among those groups who specialise in the study of culture as their object of study such as social anthropologists, and thus, it is important to be clear about the context in which the concept is being used to prevent unnecessary criticisms and misunderstandings.

Despite the difficulty involved in defining the concept of culture, we can nonetheless draw from the above discussion the idea that the culture of a people impacts on their attitudes towards the world including their environmental issues. This has implications for the ecological practices of a people. That is, differences in culture lead to different ways of addressing issues, including environmental ones, even if the approaches of different cultures are similar. Sutton and Anderson (2010: 97) emphasise that 'All people belong to a specific culture, a group of people who share the same basic but unique pattern of learned behaviour. As such, each culture has a distinct ecological adaptation' (2010: 97). The logical implication of the above discussion, however, is that as long as there are different cultures, there will be diversity in approaches to similar issues.

What I can say by way of concluding the discussion on culture is that one way to understand what culture is about, is to be guided by Geertz Clifford`s advice that:

> If you want to understand what science is, you should look in the first instance not at its theories or its findings, and certainly not at what its apologists say about it; you should look at what the practitioners of it do'(1973: 5).

By implication, to understand culture, one has to first look at what practitioners of the culture (human beings) do and not what theorists say about it. I am by this not saying that theorising culture is not useful, what I am stressing is that we learn from the known to the unknown so that starting from the objects of culture will lead to better theory about culture rather than the other way round.

However, despite the difficulty involved in defining the concept of culture, I can nonetheless draw from the above discussion the idea that the culture of a people impacts on their attitudes towards things including their environmental issues. This has implications for the ecological practices of a people. That is, differences in culture lead to different ways of addressing issues, including environmental ones, even if the approaches of different cultures are similar. Sutton and Anderson (2010: 97) emphasise that 'All people belong to a specific culture, a group of people who share the same basic but unique pattern of learned behaviour. As such, each culture has a distinct ecological adaptation' (2010: 97). The logical implication of the above discussion, however, is that as long as there different cultures, there will be diversity in approaches to similar issues.

Environmental Crisis

Park (2007: 4) notes that 'since the 1970s, scientists have been writing about what they term the "environmental crisis" and more recently this debate has evolved and broadened, incorporating some other social issues of the day into the wider *green* debate'. That is, the term resulted from the environmental consciousness that emerged in the 1960s and early 1970s (Park, 2007: 4). It was coined by the environmental activists of the 1960s and 70s to describe the catastrophic end of the world that they envisaged would occur should humans refuse to change their 'careless' attitude towards nature or the environment in general. Rachel Carson's *Silent Spring (1962) and* Lynn White's '*Historical roots of our ecological crisis'* (1967) gave much currency to the term. It is important to stress that environmental problems abound in all regions of the world, but that the global awareness of problems was created by the activities of the Western world and American environmental activists.

The environmental crisis is now seen as one of the most serious problems facing humanity. In fact, it has become a global issue that no society is completely immune from, as the threats that the crisis poses are to humanity as well as the ecosystem in general (Ojomo, 2010).

The magnitude of the crisis led Thomas Berry to pose an important question: 'Is the human a viable species on an endangered planet?' (Cited in Tucker and Grim, 2000: xxix). It is perhaps in the light of this that Park (2007: 184) remarks that, when examining the implications of the Gaia Hypothesis, 'the Earth will probably survive, no matter what humans do to it, but its survival might not include humans'.

Therefore, by 'environmental crisis in Ghana', I am referring to the poor state of Ghana`s environment due to the 'careless' way many Ghanaians have been relating to the environment, which has resulted in the loss of about two thirds of the country's forests cover, insanitary conditions, less and unreliable rainfall pattern, excessive heat with its attendant diseases and so on as the above and the following discussion will show.

One disturbing component of the environmental crisis is that as governments, environmental organisations and activists debate how to deal with the crisis, resource depletion, species extinction and pollution continue to be experienced the world over. One can also cite the destruction of the ozone layer and increase in sea levels in Mozambique and South Africa (see Mawere, 2011). According to Tucker and Grim (2000), the crisis has been compounded by population explosion, industrial growth, technological manipulation, and military proliferation. Furthermore, many accounts have shown that the crisis has affected even the basic elements that sustain life–sufficient water, clean air, and arable land (Tucker and Grim, 2000). The above problems are indicative of the nature and seriousness of the crisis that we are confronted with.

Tucker and Grim (2000) further observe that people view the environmental crisis as a complex issue, and that it is not only the result of certain economic, political, and social factors, but must also be seen as a moral and spiritual crisis that requires us to develop 'broader philosophical and religious understandings of ourselves as creatures of nature, embedded in life cycles and dependent on ecosystems' (Tucker and Grim, 2000). They argue that religions have a role to play in addressing the crisis, as they can help to shape our attitudes toward nature positively in both conscious and unconscious ways. This is because religions are full of stories that enable us to interpret the nature and destinies of humans. This,

Tucker and Grim (2000) contend, is found in the worldview of a society. It is against this background that a whole chapter (i.e. chapter 3) is devoted to exploring the worldviews of the population studied in this book. The idea of using religion to engage with the environmental crisis is not a new one, as Lynn White and others raised such suggestions long ago, and DeWitt (2000, p. 297) carried the idea further when he emphasised that 'a traditional belief in the sovereignty of God as creator leads clearly to an awareness that contemporary environmental degradation is blasphemous'. He was emphatic that:

> Authority over things belongs to the Author of those things: we have no authority to destroy what we ourselves did not create; destruction of a grand master's work by its onlooker, beholder, or curator may be a disgrace to their creator (DeWitt, ibid.).

The above discussion points to one important fact – that religion is connected to the crisis, and must also be a part of its solution (White, 1967).

Summary

The foregoing discussion has emphasised that the role of the 19th century environmental activism of the USA, Racle Carson's Silent Spring in 1962 as well as White's *The Root of Our Environmental Crisis* cannot be ingored when the origins of 'moderm' environmentalism and, particularly religious environmentalism is being discussed. This thus puts religious environmentalism into its historical context. Another important focus of the chapter is the conceptual analysis of the key terms in the book. This was aimed at assisting the reader to be clear about how these terms are used in the book.

PART II:

THE STUDY SITE AND THE PEOPLE

CHAPTER TWO

BEREKUM TRADITIONAL AREA

Introduction

This chapter outlines why the study area-Berekum Tradional Area-was selected for the case study and as well provides a brief overview of the people (of Berekum Tradional Area) on whom this book centres. Attention is also given to the geographical location, physical, local economy, political and social as well as religious organisation of the Berekum people.

The suitability of the Berekum Traditional Area as a case study

Berekum Traditional Area was selected for the following reasons:

i) Being a member of the traditional area and resident in the area for the past four decades, I have observed, over the years, negative environmental changes.
ii) It exhibits the two major types of landscape in Ghana, as it is located in the transitional zone between the savannah and the forest zones. This will enable a greater generalisation of the findings to Ghana as a whole.
iii) There are few studies on the environmental situation in Berekum despite the reality that the rate of environmental degradation in the area is quite alarming (see the study area section for details).

A preliminary survey of the literature on environmental degradation in the Berekum area revealed that there is scanty literature in this area, and I have yet to find any research that deals with religion and environment in Berekum. All that I have been able to find is scanty information on the district which appears on the internet with the title: 'Ghana District: A repository of all districts in the republic of Ghana'; little information in Blay et al. (2003: 2, 12) regarding the Pamu-Berekum Forest; and the

Municipal Assembly`s Survey, which touches tangentially on some of the environmental issues in the area. This study contends that the reason that there is so little information on the environmental situation in Berekum is grounded in the fact that sensitive areas, such as high levels of pollution, the destruction of rivers and sacred groves, poor sanitation, the high level of noise pollution in the Berekum township, the activities of chain-saw operators, and the gradual depletion of the vegetation of Berekum, among others, have been ignored in the Municipal Assembly Survey. Because of this fragmentary information on the environmental situation in the municipality, little meaningful policy has been developed to deal with the problems listed above. The study area is examined in detail in Chapter Two.

Who are the Berekum people?

The Berekum people are one of the Bono people within the larger Akan ethnic group of Ghana, and speak the Bono dialect of the Akan language. Ghana's population census put their number at 129,628 (Ghana Statistical Service, 2000). Oral tradition and fragmentary information from the few available written documents from the Berekums explain that three Akan groups migrated to and settled in the Berekum area. One of these groups was the *Awasu* people (now the people of Biadan), who were said to have migrated from Denkyira in the Central Region of Ghana to settle at Berekum through Abofo in the Offinso Area of the Ashanti Region. Berekum Traditional Council Records (1949) put this migration as having taken place between 1699 and 1702, which coincided with the period of the Denkyira-Asante wars. This migration was said to have been led by an eminent hunter called *Bɔfoakwa*, whose name literally means 'Hunter Akwa' (see Benneh, 1990; Awuah-Nyamekye, 2012b).

Another prominent hunter known as *Bɔfoɔbɛm* was also said to have led the second group of migrants from Jukwa in the Denkyira area of the Central Region of Ghana. This group of migrants settled at Abi, which is now a village that lies approximately 4 kilometres north-west of Berekum.

The third group of migrants who came to settle in Berekum comprised soldiers from Asante-Asokore. They were said to have arrived on the eve of the Asantehene`s campaign against Adinkra, the chief of the Gyamans in 1818. The Asante army was led by Sefa Antwireboa, the *Asokorehene* [the chief of Asokore] of the time. According to the sources, the Gyamans were defeated, but some of their soldiers managed to escape to the Republic of La Côte d`lvoire. As a result, the Asantes thought it wise to leave some of their soldiers behind to check on any resurgence of the defeated Gyamans. Therefore, a security post was set up at Berekum by the Asantes (see Awuah-Nyamekye, 2012b).

The above brief historical overview of the Berekums suggests that their religious and cultural practices should not fundamentally differ from other Akans in Ghana, and perhaps those outside Ghana. The Akans constitute the largest ethnic group in Ghana, and extend outside Ghana, being found in the south-eastern portion of the Republic of Ivory Coast or La Cote d'lvoire (Warren, 1986, p.7; Beeko, 2005). They are found in six out of the ten administrative regions in Ghana, viz: Brong-Ahafo, Ashanti, Eastern, Central, Western and a small part of the Volta region (see Awuah-Nyamekye, 2009a). Ghana's 2010 population census found that the Akan population comprises 47.5% of the country's total population. The Akans are culturally homogenous and, by and large, matrilineal–i.e. they inherit through the mother's line, except for the Akwapim of Larteh and the people of Mampong, who are patrilineal–i.e. they inherit through the father's line (Pobee, 1979: 44).

An Overview of Berekum Traditional Area

Berekum Traditional Area is one of the many traditional districts from the twenty-two districts in the Brong–Ahafo region of Ghana. The Berekum District Assembly was created in 1989 with the splitting of the former Berekum-Jaman District following the decentralisation policy adopted by the Government in 1988. The district has now been raised to a municipality, with Berekum as its capital, and the area comprising 33 towns and villages. The Ghanaian system of administration has zoned the country into ten regions, within which are the Districts, Municipalities and Metropolitan areas. An area is designated as a District if the population is greater than 75,000, a Municipality if the population is greater than 95,000, and a Metropolitan area if its population is greater than 250, 000. Currently, there are 216 districts in Ghana, consisting of 164 District Assemblies, 46 Municipal Assemblies and 6 Metropolitan Assemblies.

In Ghana, the name of the capital or the headquarters of a traditional area is generally used to refer to the entire traditional area. Interestingly, the names of the district, municipal and metropolitan areas have followed the same trend. Thus, Berekum, which is a town, is also the name for both the traditional area and the Municipal area. The terms Berekum Traditional Area and Berekum Municipality or Municipal area will thus be used interchangeably, as they refer to the same area.

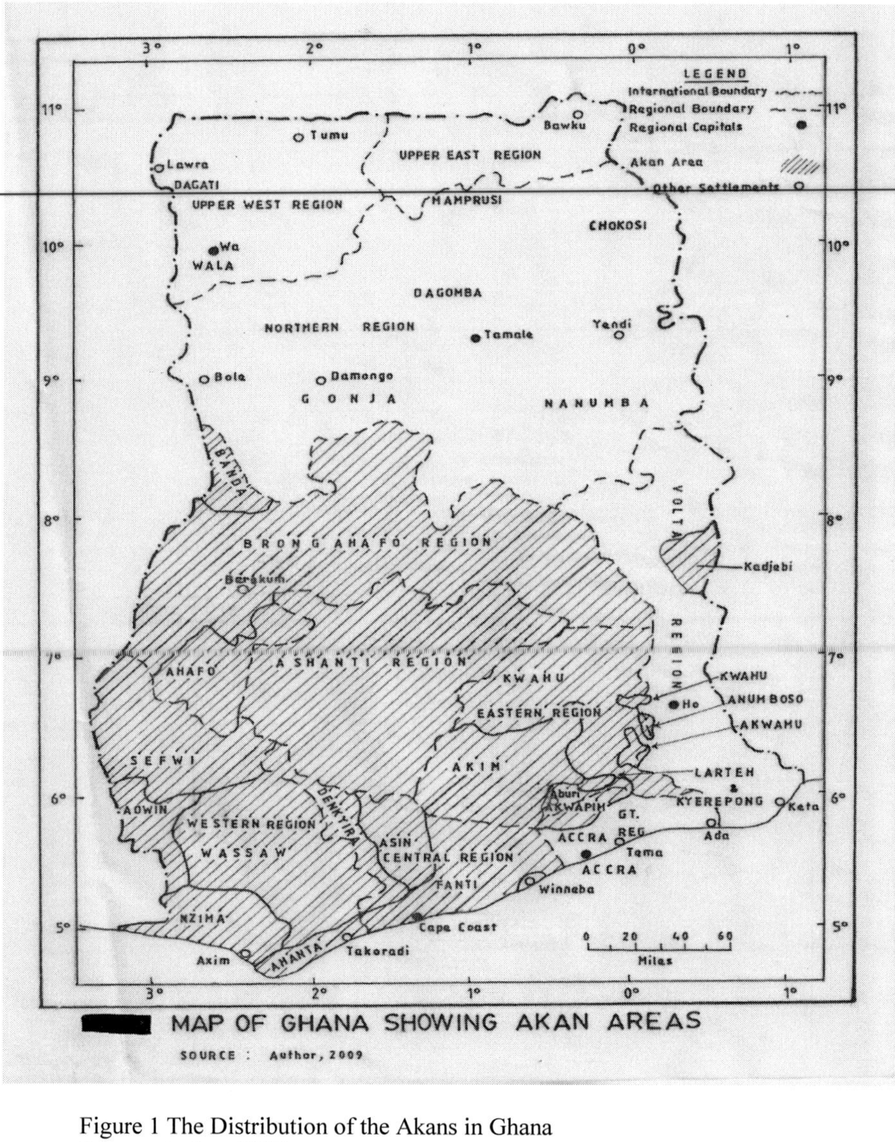

Figure 1 The Distribution of the Akans in Ghana
Source: Author, 2009

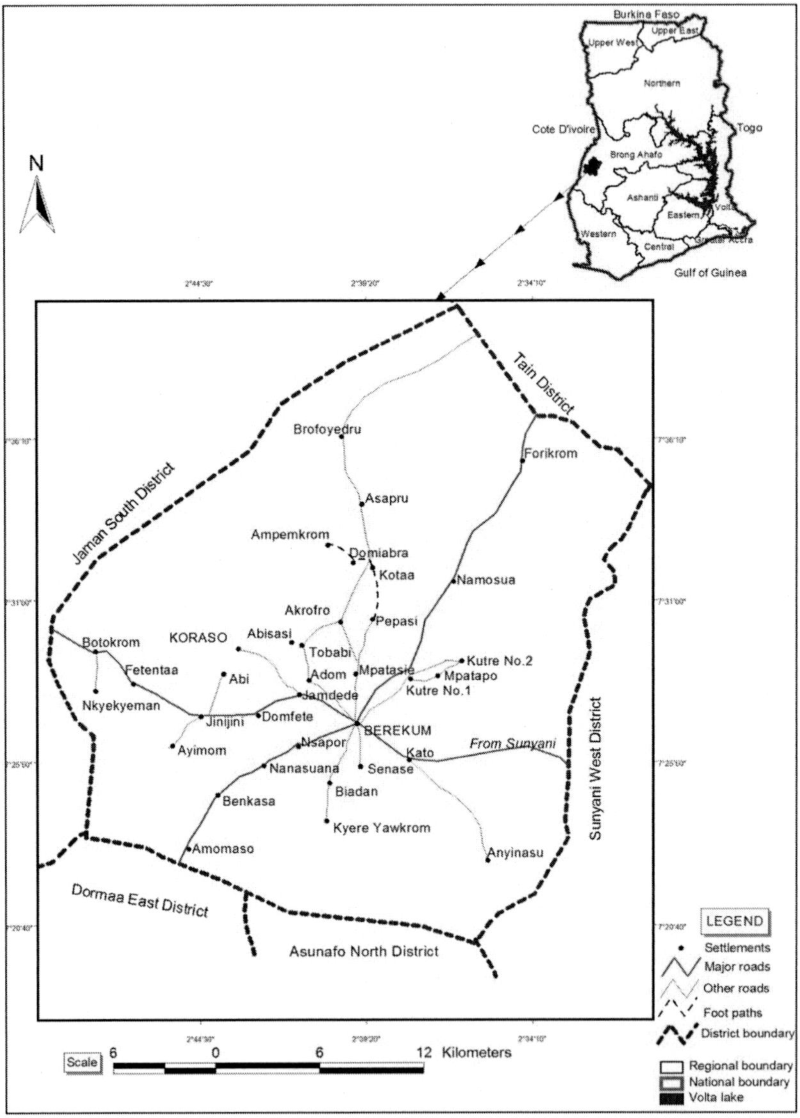

Figure 2 Map of Berekum Traditional Area–The Study Site
Source: Author

Geographical Location of Berekum

Berekum Traditional Area is located at the Western part of the Brong-Ahafo region of Ghana (see fig. 2). It lies between latitude 7`15` South and 8`00` North and longitude 2`25` East and 2`50` West. It shares boundaries with the Tain District to the Northeast, the Jaman South District to the Northwest, the Dormaa East District to the South, the Sunyani West District to the Southeast, and the Asunafo District to the South. Berekum is the capital of the Berekum Municipality, which has a total area of 1094.2 square kilometres. The population of the Municipality is 129,628 (Ghana Statistical Service, 2010). The growth rate is estimated at 2.5% per annum (Berekum Municipal Heath Annual Report, 2007). Berekum is 32 km from Sunyani, the regional capital, and 437 km from Accra, the national capital.

Physical Characteristics of the Area (relief and drainage)

The Berekum Traditional Area is 800-900m above sea level (Berekum Municipal Health Report, 2007), with a landscape of undulating landforms that are interspersed by isolated hills. Its rivers are small and the area is generally well drained, hence serious flooding is uncommon. The area falls within the moist semi-equatorial climatic zone and has a double maxima rainfall regime, with a mean annual rainfall of between 1270-1524mm (KOB Consulting Ltd., 1988: 17). The first and heaviest rainfalls occur in June, with the second rainfalls being from September to October every year. The dry season, which is referred to as *Harmattan*, takes place from December to March, during which time the trees shed their leaves and generally appear brown. The original vegetation in the area was mostly moist semi-deciduous forest, but this has been severely depleted through a combination of inappropriate farming methods, indiscriminate logging, rampant annual bush fires, and a general disregard for indigenous conservation models, with the result that grass species are gradually taking over the vegetation of Berekum Area.

The Local Economy

The physical characteristics of Berekum favour an agrarian economy, with farming being the major economic activity in the area. Many kinds of food crops are cultivated in Berekum, but the major ones are cocoyam, plantain, yam, cassava and maize. Vegetables such as okra, onions, garden eggs, pumpkins, pepper and tomatoes are cultivated as well.

The major cash crops of the area include cocoa, cashew nuts and oil palm, and Berekum is well known for its cocoa production. According to Benneh (1990), Berekum ranked among the chief cocoa producing centres of West Asante in 1936, but the production of cocoa became so intensive that the existing forest was quickly depleted. This compelled the farmers to migrate to Sefwi, in the Western Region of Ghana, to cultivate cocoa. The dominant method of farming is traditional, utilising cutlasses and hoes for clearing the bush and harvesting the produce. Animal rearing is also important in the study area, where the people also engage in hunting that involves the uses of local guns produced by the blacksmithing centres in the area. Lumbering is also practised, and there is currently one large sawmill in the area that processes wood and plywood for local consumption and for export. There are an uncountable number of what have become known as 'table saw-mills', which are dotted around the residential parts of the study area.

The Berekum Municipal Assembly report of 1996 says that about 87.7% of the citizens of Berekum are employed, and the report breaks down the local economy as follows:

Table 1: Breakdown of local economy

No	Sector	Percentage (%)
1	Farming	67
2	Industry	16
3	Services	17

Source, Berekum Municipal Assembly Report, 1996

However, according to ghanadistricts.com (12 November 2010), the farming sector now only makes up 57% of the workforce. This could mean that both the services and industry sectors have benefited from this drop, but the report was silent on this matter. It is clear that the greatest percentage of the workforce in the study area is in the farming sector, which means there is pressure on the land in the traditional area.

Political Organisation

Berekum uses the chieftaincy system of governance, and Chiefs are selected through maternal lineage.[1] A chief in Berekum is usually considered as the first citizen of the village, town or state. He is viewed as the source of all traditional authority, because he is regarded as representing the founding fathers of the state. This makes the office of the chief a sacred one, as he is regarded as the earthly representative of the ancestors (Busia, 1968; Boafo-Arthur, 2001; Awuah-Nyamekye, 2009c, 2012; Odotei and Awedoba, 2006; Owusu, 2006). The appointment of a chief is normally preceded by divination and other rituals, such as the offering of prayers and sacrifices, which are meant to help the kingmakers in selecting a candidate that is acceptable to the people and the ancestors. Hence, the chief is accorded the greatest respect and obedience in traditional society. The Berekum Traditional Area was given paramountcy status by the colonial government in 1901. This was the alleged reward for the loyalty that the people of Berekum demonstrated to the British during the Asante uprising of 1900-01 (Arhin, 1973).

As in other Akan traditional societies, a chief in Berekum has staff that assists him in the day-to-day administration of the area. The highest traditional authority is vested in the *Ɔmanhene* (Paramount Chief), who is the supreme ruler of the traditional area. He is followed by the *Ɔhemaa* (queen-mother), then the *Asafohene* (wing or divisional chiefs), who are followed by the *Adikro* (town or village chiefs). The next in line are the clan or lineage heads, with the household heads occupying the bottom space in the hierarchy. The following table constructed from my interviews shows the main traditional political offices in their hierarchical order in the Berekum Traditional Area, and the roles involved therein:

[1] It is interesting to note that in Akan traditional societies, a woman can become a chief in some communities although currently chieftaincy is a male dominated office in Ghana. Historically the reverse was the case (see Meyerowitz, 1958).

Table 2. The traditional administrative structure in Berekum Traditional Area

Hierarchy	Role
Ɔmanhene	The overall or the supreme ruler of the entire traditional area
Ɔhemaa	The counterpart of the *Ɔmanhene* who is in charge of matters of women in the traditional area
Kurontihene	The *Kurontihene* assumes responsibility of the traditional area/town in the absence of the *Ɔmanhene*
Akwamuhene	The *Akwamuhene* is second to the *Kurontihene* in the traditional administrative structure of Berekum
Twafoɔhene	He acts as the spokes person for the sub-chiefs in the area. Traditionally he is in charge of designing the path for the traditional army in times of war
Adɔntinhene	The *Adɔntinhene* leads the army during war
Nifahene	*Nifahene* administers the right quarter of the traditional area on behalf of the *Ɔmanhene*. He is also the chief in charge of the right wing of the moving army.
Benkumhene	Just as the *Nifahene*, the *Benkumhene* administers the left quarter of the traditional area on behalf of the *Ɔmanhene*. He is also the chief in charge of the right wing of the moving army.
Kyidomhene	*Kyidomhene* serves as the rear guard of the moving army
Ankobeahene	The main duty of the *Ankobeahene* is to remain home to protect women and children whiles the army is away in a battle
Gyasehene	The chief in charge of the royal household (palace) and the welfare of immigrants in the area
Sanaahene	This is a chief in charge of treasury of the *Ɔmanhene*
Akyempimhene	The chief of the sons and daughters of the *Ɔmanhene*
Mawerehene	*Mawerehene* is a chief is in charge of the security and the welfare of the family members of the *Ɔmanhene*
Nkosoɔhene	*Nkoso* means development. Therefore, the main role of *Nkosoɔhene* is to take charge of issues that border on development in the area.
Apagyahene	He is in charge of the regalia of the substantive chief of a place. He makes sure that the chiefs appear neat in accordance with the customs and traditions of the place.

Source: Author's fieldwork notes

The above structure is replicated in the villages, where the *Ɔhene* (chief) is the head, since there should be only one *Ɔmanhene* in any traditional area.

There are currently two sets of political administration in the Berekum Traditional Area; the central government being responsible for its day-to-day running through the Municipal Chief Executive (MCE) of Berekum, whilst traditional matters (such as enstoolment and destoolment of chiefs, stools lands, civil disputes etc) are handled by the *Ɔmanhene.* However, it is important to note that the real political power in the area belongs to the Municipal Chief Executive (central government), as the MCE can take certain decisions without any consultation with the traditional head of the area–the *Ɔmanhene*. For instance, the MCE presides over security matters in the Municipality or traditional area and can ban any activity in the area–including that of the *Ɔmanhene* and his elders–that he deems to endanger the security of the area. For example, the MCE can ban the celebration of a local traditional festival if he/she has a reason to do so. At times, this can happen for political expediency, for example, if the MCE believes that the *Ɔmanhene* is politically aligned with the opposition party in the country.

Theoretically, all the pre- and post-independence constitutions of Ghana recognise the authority of chiefs (Article 13 of the 1960 Constitution of Ghana stipulates that 'chieftaincy should be guaranteed and preserved' and Article 270 of the 1992 Constitution of Ghana prevents parliament from enacting any law that directly or indirectly intends to control the institution of chieftaincy), but this does not transfer into practice. The above development means that the chiefs no longer have total control over issues affecting their people and the environment.

Social Organisation (Kinship ties)

I have already pointed out that the Berekum people are one of the Bono people forming part of the larger Akan ethnic group, and speak the Bono dialect of the Akan language. Socially, the people of Berekum are hierarchically organised. At the top of the social ladder is the ruling class, that is, the *Ɔmanhene* and other members of the royalty. The divisional chiefs and the sub-chiefs follow them and next come the commoners or the free-born. Some time ago, the bottom of the social ladder was occupied by domestic slaves, who were part and parcel of their masters' households, with those judged to be 'good' and hardworking sometimes even being allowed to marry from their masters' households. Later on, when slavery was completely abolished, they were integrated into their owners' families (Awuah-Nyamekye, 2012). The basic social unit in the Berekum society–

as it is in all the Akan societies–is the matrilineal family (Rattray, 1923, 1959; Danquah, 1928; Nkansa-Kyeremanteng, 1999).

However, due to the unprecedented process of cultural interaction and evolution, the extended family–the glue that holds members together–has been seriously weakened. The factors responsible for this have been Western education and culture–which encourage the nuclear family system--and the demands of the modern economic system (where people are separated from their home communities areas due to work), among others. The only time at which the collective responsibility associated with the extended family system remains strong is with the death of family members.

The foregoing provides a summary of the basics of Berekum society. The following section presents the area's religious demography, which plays a major role in environmental issues in the Berekum societythere.

Religion

Religion is a notable aspect of Berekum culture, being infused in everyday lives, with secular aspects of lives being hard to separate from spiritual ones. Prior to their encounter with the Western world in the latter part of the 19th Century, the Berekum indigenous religion–which is now known as African Traditional Religion–was the only religion known to them. This faith has profoundly influenced their lives and thought. Berekum people believe in a host of spirit beings, with *Onyame* or *Onyankopon* (God) being seen as the creator, controller and sustainer of the world. Many of the other spirit beings are believed to inhabit images, rocks, mountains, caves, trees, rivers, and other natural objects [temporary], and this belief made the people treat such natural objects with a form of reverential respect. However, in the present day, due to the people's contact with the outside world through colonisation, Western education, foreign religions and other meaning-giving systems, the traditional area is religiously plural. This pluralism is further discussed in Chapter Four of this book .

Summary

The discussion has not only been on the reasons which informed the selection of the Berekum Traditional Area as a case study but has also dealt with who the Berekums are. Particular attention has been given to issues such as the geographical location and the physical environment of the area. Special attention is also given to how the Berekums organised themselves socially, politically and religiously.

CHAPTER THREE

THE CURRENT ENVIRONMENTAL SITUATION IN THE BEREKUM AREA

Introduction

In the preceding chapter an overview of the study with particular reference to the geographical location, physical, local economy, political, social as well as religious organisation of Berekum Area have been discussed. This chapter makes a survey of the current environmental situation in the study area and as well discusses issues such as the factors responsible for the current environmental situation and the effects or impacts of the problems on the people in the study area in particular, and in Ghana in general.

The current environmental situation in the Berekum area

> The land is now bare and degraded. Almost all our forests are gone and several plants and animal species are no longer here. It was only in a few areas that one could see grasses. But now the place is gradually becoming grassland and most streams and water bodies are drying up (Akua Amponsah, personal communication, 7 October 2011).

The above quote is the answer that Akua Amponsah, a farmer in Berekum, gave when I asked her views on the current environmental situation in the Berekum Traditional Area, and epitomises the environmental reality of Berekum today. The details are examined in the following paragraphs.

Deforestation

Deforestation, according to the United Nations Conference on Environment and Development (UNCED), is the 'land degradation in arid, semi-arid, and sub-humid areas resulting from various factors including climatic variations and human activities' (UNCED, 1992). The conference also explained that deforestation should not be confused with degradation, stating that while deforestation reduces the size of forests, degradation involves the decrease in the condition (quality) of the forest, which manifests itself in the vegetation, faunal and floral diversity, the fertility of the soil, and the other components of the ecosystem of forests (UNCED, 1992). The distinction between deforestation and degradation is important, as the first does not necessarily imply the second and vice versa. However, in the case of the study area, both have occurred.

The rate of environmental degradation and deforestation in Ghana has already been noted (see p. 3). Various arms of government of Ghana have made several attempts to deal with the situation by developing plans to ensure the sustainable management of the harvesting of forest products (MLF, 1994, 1996; Donkor and Vloski, 2003). As part of its effort to protect and conserve the forests in Ghana, the government of Ghana has created a total of 266 National Forest Reserves in the high forest zone, which occupy 1,634,100 hectares of land (Hawthorne and Abu-Juam, 1995 quoted in Donkor and Vloski, 2003). Berekum Traditional Area has two forest reserves–Tain II and Pamu-Berekum. However, the exploitation of the forest in these reserves continues despite its illegality and the government's efforts to stop it. It is sad to note that these national assets are diminishing seriously in both size and quality.

There is evidence of theft of high value trees by both timber contractors and chain-saw operators in the area. It was reported that some corrupt forestry officials collude in these illegal activities. Many of my informants said that timber contractors and chain-saw operators indiscriminately cut the trees, including the endangered ones. I was told in the villages close to the forests that they hear the noise made by chain-saws deep into the night on an almost daily basis. The following statement taken from the Tropical Timber Report (TTR) in 1990, which is attributed to Mr. Joe Ackah–Brong Ahafo Regional Zonal Manager for Ghana's Forestry Commission (FC) – attests to the stealing of timber in the study area:

> [..] about ten uncertified mobile sawmills were identified in Tein II and Paamu/Berekum forest reserves. Authorities were able to impound over 1,200 teak billets now placed at the Berekum District office of the Forestry Services Division. The billets were meant for export and could have costed [sic] the government several millions of cedis. Mr. Ackah said that illegality occurred due to [a] lack of logistics and capacity on the part of the FC (Forestry Commission) and [a] lack of cooperation among the police, the community and the FC. (TTR, 2006, p. 2).

In providing some background to the problem, Mr. Gyedu Augustine observed that the increase in illegal logging is due to the inability of the Forestry Commission to check timber felling and to ensure concessionaires comply with harvesting procedures. The local people are of the view that the government's redeployment of forest guards in Ghana has been one of the major contributory factors since the presence of the guards scared the illegal timber contractors, particularly the chain-saw operators.

According to the local people, the Tain II and Pamu-Berekum Reserves are now shadows of their formal selves. This is corroborated by Mrs. Philomena Appiah Boakye, the Regional Director of the Environmental Protection Agency (EPA) for the Brong-Ahafo Region. In her Keynote address to the public during the 2011 World Environmental Day, held at Kato near Berekum, she revealed that:

> The Pamu-Berekum Forest Reserve was established in 1932 with a total area of 189 km^2. The forest area in 1990 had been reduced to 91 km^2 (a loss of 98 km^2 in 58 years); in 2000 it had been reduced to 1 km^2 (a loss of 90 km^2 in just 10 years). In the other instance, the Tain II Forest Reserve was established in 1934 with a total area of 509 km^2. The forest area in 1990 had been reduced to 451 km^2 (a loss of 58 km^2 in 56 years); and in 2000 it had been reduced to just 108 km^2 (a loss of 343 km^2 in just 10 years) (Appiah Boakye, 2011).

Mr. Gyedu (personal communication, 13 December 2012, Berekum), a forester from Berekum, explained that although the actual reserve (land) is intact, the tree cover is reduced. This is environmentally unhealthy, since increasing the number of trees is believed to be the answer to counteracting global warming (Burley and Haslam, 2008). The concern is that if even government-protected areas are being degraded in this way, then the fate of non-protected areas in the study area is all the more certain. A report by Forestry Outlook Study for Africa (FOSA) and the Ghanaian Ministry of Lands and Forestry has the following to say on the consumption of forest products:

> A major problem in Ghana is the high consumption of wood for energy and cooking, estimated at 14 million m^3 annually and expected to increase to 20 million m^3 in 2010. Even in urban areas, about 69% of all households use charcoal. Agyarko found that 91% of total roundwood production is used as fuelwood and for charcoal. The remaining (9%) is used as industrial roundwood (mainly timber) (FOSA, 2002).

It is thus unsurprising that a recent report has indicated that there has been a significant rise in Green House Gas (GHG) emissions in Ghana (Edjekumhene et al., 2011), and this now puts the country into the bracket of being a net-*emitter* of GHGs. The above statistics show the level of degradation in the study area, and in contemporary Ghana more generally. Notably, it is the vegetation, especially the trees, that are most affected. Another illegal practice closely linked to illegal logging is the illegal poaching practised by hunters in the forest.

Poaching

Some of my informants (especially those who live close to these forests) discussed the illegal activities of poachers, who secretly enter the reserves to hunt for game with traps. It was revealed that some corrupt District Forestry Officials connive with the hunters, and this affects the protection of endangered species in the traditional area. The hunters I interviewed generally agreed that their activities had adversely affected the faunal capacity of the area, and some of the animal species they mentioned hunting have been considerably reduced in number, including the groove-toothed forest mouse (*Leimacomys buettneri*); the toothed shrew (*Crocidura foxi*); the red-fronted gazelle (*Gazella rufifrons*); the spotted wild dog (*Lycaon pictus*); the white-thighed black-and-white colobus (*Colobus vellerosus*); the white-necked otter (*Lutra maculicollis)*; the pied bat (*Chalinolobus superbus*); the bongo (*Tragelaphus euryceros isaacii*); the bush buck (*Tragelaphus sylvaticus*); the buffalo (*Syncerus caffer*); the black duiker (*Cephalophus niger*); the side-striped jackal (*Canis adustus*); and the grey parrot (*Psittacus erithacus*)–the final three having now completely disappeared from the area. Many of these species are also listed as globally threatened species by the International Union for the Conservation of Nature (IUCN).

Studies have established that hunting for game in West and Central Africa poses a great danger not only to biological diversity and ecological processes, but also to future generations of people, whose protein needs and income are highly dependent on bushmeat (Bowen-Jones et al., 2002 and Milner-Gulland et al., 2003, cited in Swensson, 2005: 5). The FAO Twenty-Third Regional Conference for Africa (2004)–which had the theme 'The bushmeat crisis in Africa: Conciliating food security and biodiversity conservation in the continent'–noted that 'the unsustainable bushmeat utilization in West and Central Africa is currently one of the most important food security and biodiversity conservation challenges facing the region'.

In Ghana, a ban is usually placed on the hunting of all wild animals with the exception of the grass cutter during 'close season', which takes place between August and December. *Ɔpanin* Yaw Barima an experienced hunter at Mpatasie, told me that the ban is a pre-colonial practice, but has been adopted by the governments of Ghana (Personal Communication 18, November 2011),. It became a regulatory mechanism backed by the Wildlife Conservation Regulations L. I. 685 of 1961. However, despite this regulation, it is common to see bushmeat being sold by 'Chop bar' operators (local restaurants) in Berekum, and indeed throughout the country all year round.

Bushfires

According to my key informants, the most serious of all the factors that impact on the environment are the annual bushfires the traditional area experiences, especially during the *Harmattan* (dry) season. It was also mentioned that bushfires are the principal threat to sacred groves in the area. The local people said that several decades ago, all the towns and villages in the traditional area used to have sacred groves. In some cases, one could count as many as ten groves in one village or town. However, most of these groves have now been degraded or destroyed through bush fires. Bushfires generally degrade forests since they cause forests to lose their natural status and their ability to influence the weather in the area My interviewees also said they were partly responsible for the loss of wildlife in the area. This was noted to have far reaching ecological implications. For instance, some of the experienced farmers I interviewed pointed out that many plants, and animals such as worms and insects, play crucial ecological roles, including the recycling of nutrients, and ultimately aid soil fertility. These empirical claims have been ratified by scientific studies elsewhere (see Fisher, 1978; Morris et al., 1991).

In addition to the loss of land fertility, bushfires also lead to the loss of important medicinal herbs, which a preponderance of the Berekum people rely on for their health needs (see Brown, 1992). The above situation partly explains why species of grass are gradually taking over the vegetation that comprises the study area. Interestingly, some respondents said that the emergence of grasses in the area is now encouraging the rearing of cattle there. In view of this development, the presence of the Fulani (major cattle farmers in West Africa) is on the increase in Berekum. However, there is strong resistance to this development, and there have been a number of protests due to the destruction of people's farms. I was told by Ameyaw Martin of Akroforo (Personal Communication 13,

November 2011) that these cattle owners have secretly been sending their herds to graze in the national reserve. Bushfires are sometimes accidental, but the phenomenon of group hunting is emerging fast in the study area and, according to one hunter that I interviewed–Kwaku Amoabeng, (Personal Communication, 13 December 2011)–hunters deliberately set fires during the *Harmattan* (dry) season in order to force game out to be shot (or else killed by the fire). My informant said that this practice represents a new phenomenon in the area, and that although there is legislation against starting bushfires, its enforcement is very weak. Mr. Gyedu (Personal Communication, 16 October 2011) explained that bush fires are a central cause for the rapid depletion in forest cover in the Berekum Area.

Threat to the Koraa River and its Fish Sanctuary

Berekum Area does not have big rivers, but River Koraa is the principal river in the study area, which supplies water for domestic and industrial use in the entire area. This river is now under serious threat. Almost every interviewee pointed out that there had been a heavy encroachment on the Koraa basin and this has led the once-forested basin of the river to be gradually taken over by grasses. There used to be a fish sanctuary on the river, which the local people (and tourists to the area) visited to watch fish in the river from the bridge on *nkyida* (sacred days or holidays). Now, however, the *Asuo Koraa* (river Koraa) has become seasonal, and it is only during the sporadic rainfalls that the fish come out. At the time I was gathering the data, weeds (mostly grasses) had taken over most of the section where people stood to watch the fish from the bridge, and it is now difficult to see the fish even when it rains.

Figure 3 The Koraa fish sanctuary now

During the data collection phase of my research, I found out hat land very close to River Koraa–the main river in the study area–had been sold, and is now under plantain and coconut plantation. Nana Adjei informed me that many rivers are being destroyed by the activities of sand contractors, who go very close to the river-beds to win sand. The trenches they leave after their activities impede the smooth flow of the rivers because they collect some of the water from the rivers and this affects their natural flows. When the local people complained about this, the contractors claimed that they have been issued permits by the Municipal Assembly of Berekum (Personal Communication, 17 November 2011). It was also noted that another river–the River Berekum, from which the traditional headquarters 'Berekum' takes its name–is almost dried up now. This is because its catchment area is now almost covered with buildings. Some of these buildings are virtually in the channel of the river, as are farms close to the banks of the river. It surprises many people that the catchment area for this river houses an important sacred grove in Berekum, within which important rituals connected with the main festival of Berekum (*Kwafie*) are performed. Nana Kwabena Wusu (Personal Communication, 8 September 2011) told me that all the blackened stools of the past chiefs of Berekum are brought here during the evening of the festival day for ritual bathing (washing). A car-washing bay has now even been constructed close to the river and the waste water flows into the river.

Pollution

According to Mr. Oliver, the Berekum Municipal Environmental Officer (Personal Communication, 14 December 2011), the general state of the sanitation in the municipality is a source of worry to many people. The management of both the solid and liquid wastes in the traditional area leaves much to be desired. Heaps of refuse are also a common sight in the traditional area, especially in the Municipal capital – Berekum. One can sometimes witness stray animals in the middle of the town feeding on the refuse dumps. Moreover, the majority of people in the municipality rely on public places of conveniences, but these places are managed very poorly. The Municipal Assembly that owns

Figure 4 A heap of refuse, dumped right in the centre of the Municipal capital.

Figure 5 Caked human excreta at the dump site in Berekum

them is unable to regularly buy the needed detergents to clean the toilets. This is a major source of worry to those who live close to these toilets, because of their unbearable stenche and the attendant health problems.

The waste disposal sites in the area are also a problem as they are not efficiently managed. Human excreta is not treated, but simply dumped raw at the sites, and passers-by can be seen covering their noses whenever they are close to the dumping sites. These sites become dry and caked during the dry season, and I was told by the Municipal Environmental Officer that the caked excreta is set on fire in the dry season. This practice is another source of bush fires. All these problems are a result of inadequate resources to acquire the necessary logistics to manage them ecologically. This pollution problem impedes the fight against global warming that the world has on its hands today.

Another source of pollution is noise-making. Noise pollution is high, particularly in the Municipal capital, Berekum. This is perpetrated by drinking bars (local pubs), restaurants, music shop operators and churches, particularly the penteco-charismatic group.

The religious demography table of Berekum in Chapter Four shows that Christian-related groups dominate. In Ghana, Christians are generally noted for making high levels of noise during their worship sessions as the use of modern musical gadgets is a key element to their worship today, so even a church with a low number in its congregation may make a lot of noise with their musical instruments. At times, the level of the noise becomes unbearable during their special prayer sessions–which they called "all-night" or "half-night" sessions. Many of my informants said that people are, however, afraid to complain for the fear of being tagged as 'anti-Christ' by Christians. This makes it difficult for people to get good sleep after a day's work. There are provisions in the country's law against noise pollution, but the police are not keen to enforce these laws. Mr. Oliver, the Municipal environmental Officer (Personal Communication, 14 December 2011), lamented about this situation, but said his hands were tied since the police were not keen to enforce the law against excessive noise making in the area.

Yet another source of pollution comes from the activities of what is known as 'table-saw-milling'. Instead of setting up their milling machines at the light industrial area provided by the Municipal Assembly, plants are set up within the residential areas all over the municipality, and the dust that is produced by the 'table-saw-milling' is a source of worry to those who live close to them.

Figure 6 Some of the Table–Saw-Milling sites at the heart of residential areas in Berekum

During my fieldwork, I found out that apart from financial motive for table-saw-milling, some of the millers are completely ignorant about the effects of their industry in the area. For instance, in responding to questions about the negative effects of the 'table-saw-milling' on the environment, a table saw-miller and wood seller asked whether it made any difference if the trees were logged, since crops do well even in the northern parts of Ghana, where the vegetation is dominantly grasses with very few trees. This represents a clear confirmation of one of the concerns shared by most of my key informants–that many of the younger generations lack even basic knowledge of their surroundings, with many failing to understand the consequences of their actions, and only taking an interest in the monetary benefits that they can get. This is one of the hurdles that must be crossed in the people's quest to find a lasting solution to the environmental problems that are now confronting the area. Thus, the study area has many environmental problems, but those provided above constitute the major ones. Having surveyed the environmental problems in the study area, my next focus is to examine some of the major causes of these problems within the study area.

Factors accounting for the environmental degradation and deforestation in Berekum

From the data gathered during the fieldwork, I observed that a combination of factors is responsible for the environmental degradation and deforestation in Berekum. The main ones are discussed in the following paragraphs.

Dominance of the scientific methods of environmental management

I observed from the data that combined factors account for the current environmental situation in both Berekum and Ghana as a whole, but the issues involved will be better explained or appreciated if they are analysed within the broader context of the colonial experience and its attendant effects. My interactions with the key informants revealed that the heavy reliance on scientific methods now (by those in charge of environmental management in the area) for addressing environmental problems to the neglect of indigenous methods is one of the key problems that underwrites the difficulties they experience in trying to handle environmental issues in Berekum, and that this can be more broadly generalised to Ghana as a whole. The issue of how to balance economic growth against environmental concerns was also identified as a significant problem, and this difficulty is grounded in the fact that economic considerations are infused in all areas of Ghana's political decision-making (see Drazen, 2000).

There is an interplay between political and economic decisions with regard to resource management and how economic decisions impact on environmental issues in Ghana. Nana Akumfi Ameyaw emphasised theview that the colonial masters were to be blamed for the over-exploitation of Ghana's natural resources, particularly its minerals and forest products. He argued that the commercialisation of minerals (gold, diamonds and bauxite) and timber became more pronounced in Ghana during the colonial era. This was the time when these resources began to be exported in large quantities to the Western World (Personal Communication, 12 September 2011). Cecile Jackson (1994) has expressed a similar view: 'Early colonial policies were concerned with securing natural resources for European interests and introduced many environmental policies to achieve this end and [to] regulate usage of natural resources' (cited in Amanor, 2001, p. 64). Nana Yaw Mensah (personal communication, 7 November 2011) observed that the imposition of the colonial masters' way of life on the African negatively affected the indigenous peoples in many ways, especially with regards to indigenous ways of conserving nature. In fact, this opinion was expressed by almost all of my respondents. Akyeampong captures the influence on colonisation in the following words:

> Colonial rule in Africa privileged Western Knowledge Systems and discredited Indigenous Knowledge Systems, and provided the context for the fashioning of 'scientific knowledge' about Africa (Akyeampong, 2007).

Interestingly, some of my interlocutors mentioned that the modern National Forest Reserves in Ghana are a positive legacy of colonialism. For instance, reference was made to the creation of the Tain II and Pamu-Berekum Reserves, which are all located in the study area. A concern, however, was that during colonial times, Ghana experienced a lot of planned development, but lacked the recources to systematically analyse the potential environmental impacts of this (Adarkwa, 1983; Ofori, 1991). This trend continued after the country's independence, being guided by modernisation theory, which rejects most features of indigenous life-styles in favour of alien or modern systems (Appiah-Opoku and Mulamoottil, 1997). According to my key informants, this has had an adverse effect on the weal of the country, particularly the local culture.

Many of my interlocutors (particularly the chiefs) pointed out that the only time that politicians in Ghana make favourable comments about local people (chiefs and their subjects) is during election campaigns, when the politicians go to the various communities to solicit votes. Nana Agyemang (personal communication, 14 October 2011), for instance, observed, 'during this time one would hear that the chiefs are the fulcrum for development and their views are important for development. But, contrary to this, the politicians mostly fail to consult us, even when some developmental projects that fall within our jurisdiction are being carried out'. Corroborating Nana Agyemang's claim, Nana Yaw Nkrumah (Personal Communication, 11 November 2011) added:

> For instance, when you have an informal chat with them [politicians], you realise they recognise and appreciate the potentials of indigenous knowledge. They tell you that *nananom* have great wisdom to deal with the pressing issues in their communities most especially those that border on the management of natural resources and the conservation of the environment in general. Personally, I see this as rhetoric, for as soon as they assume political power, they relegate these ideas to the background in favour of modern scientific ones which have been inadequate in addressing the problems on the ground and, in some cases, are not applicable here.

A good number of my respondents, especially the chiefs, were not happy with the attitudes shown by contemporary Ghanaian policy-makers through their retention of many of the economic policies bequeathed to Ghana by the colonialists. For instance, economic considerations remain higher than those of environmental conservation in all the major policies in country, and logging continues to increase at an alarming rate. Mineral deposits are being overexploited throughout the country, and even lands close to river-beds/channels have been sold as building plots or farmland.

According to my respondents, one other major effect of this activity is that almost all the quality sand resources in the Berekum Area have now been used up. According to Nana Kwadwo Buadu–the Secretary of the Friends of the Earth, a local environmental NGO–Berekum people now have to buy and import quality sand from far away in the Kyiraa area of the Sunyani Municipality in order to build their houses (Personal Communication, 12 October 2011), and this comes with a serious financial cost.

Overpopulation

An important factor relating to environmental degradation that was noted by almost all my key informants is the population explosion that has occurred in Berekum. They conjectured that when the population of the area was much lower, there was little pressure on the land and other natural resources, so their management was not an issue, but as the population increase, it became much more of one. The Ghanaian demographic statistics below show this dramatic population increase.

Table 3: Size of Berekum Population in relation to Ghana (%)

Year	Berekum(District)	National(Ghana)
1960	39, 374	6,126, 185
1970	56,046	8, 579 313
1984	78.604	12, 296,081
2000	93,235	18, 922 079
2004	106,391	-
2006	113,650	-

Source: 2000 Population and Housing Census

The 2010 Population and Housing Census put the population of Berekum at 129,628. This means that it has witnessed a population increase of 36,293 in a decade. This analysis shows the increase in intensity of land use and the pressure put on the land for both settlement and farming purposes. As a result, the *Kontonkori* and *Kwaabaa* sacred groves in Mpatasie and Senase respectively and the *adwamɛ* in Biadan are now residential areas.

Another effect of this population explosion is that most of the youth move to urban centres for work. According to *Ɔkyeame* Okra Acheampong (Personal Communication, 12 January 2012), this rural-urban drift alienates the youth from their roots as city lifestyles are different, and thus most of the youth return home with worldviews that are very different to those held within the local area. The youth, he said, are those who have received Western education, and often view the traditional way of life as old-fashioned, and are unprepared to listen to the elders. They consider most of the traditional explanations to issues, including ecological views, as unscientific. Moreover, when they return to their villages, they transfer their newly-acquired worldviews to their peers in the area, who venerate them because of the wealth they have acquired from the cities. In fact, this view was corroborated by many of my informants.

Poor farming methods

Many of the farmers that I interacted with during the study identified poor farming methods as one of the major reasons for the fast depletion of the forest and the degradation of the environment, with shifting cultivation techniques being singled out. They described shifting cultivation as a method of farming in which the farmer repeats the cropping on the same land for two to four years before moving to another plot of land to allow the previous land to fallow for five to ten years before he/she returns to it again (see Akinsanmi, 1975, pp. 72f). The farmers I interviewed accepted that some years of fallow are needed for the land to regain its fertility, but contended that once the original land has been cultivated it will never return to its original state. They were unanimous in holding that although this type of farming was possible some time ago, it is no longer practicable today due to the population explosion in the area (see table 3 above). However, it is still the major farming method used in the Berekum Area–hence the high rates of deforestation and degradation there. The Chiefs, as custodians of the land, said that they wished they could do something about this, but contended that their hands were tied due to the population explosion, adding that although custom places them as the custodians of the land, the loyalty of the people is gradually being transferred from them towards the central government.

Furthermore, Yaw Barima (Personal Communication, 18 October 2011) stressed that a sizable proportion of the land is used for maize cultivation, but his experience as a farmer has shown that repeated cultivation of maize on the same land induces grass growth. He contends that this is what is responsible for the gradual replacement of the vegetation by grasses.

Protracted chieftaincy problem in Berekum

Although my respondents agreed that the influence of chiefs is waning, and they stressed that no matter what the situation, the role of the Chiefs, as agents of development in their respective communities cannot be overemphasised. According to my informants, the chiefs provide security in terms of ensuring the smooth functioning of labour and protecting the rights of the people that undertake communal labour for the execution of development projects. This explains why development often stalls in the areas where there are no substantive chiefs, or where there are protracted chieftaincy disputes. This even affects the smooth running of local governmental administration (see Clarke-Ekrong, 1977).

In Ghana, the land for development projects must be released by the chief, who is traditionally the custodian of the stool land in his/her area of jurisdiction (Busia, 1968; Yanka, 1995; Owusu, 2006; Awuah-Nyamekye, 2009c). Berekum, however, has not had a substantive *Ɔmanhene* for more than a decade due to a protracted chieftaincy dispute that erupted after the demise of the substantive *Ɔmanhene*, Nana Owusu Yiadom II. This has made the situation worse in Berekum, and the chiefs I interviewed did not hide their frustration over the matter. This issue of chieftaincy is revisited in Chapter Seven where the special role that this institution plays in the life of the people is analysed.

Apathy

My respondents stressed the impact that people's apathy towards government property is having on development in the area, and indeed throughout Ghana. Several theories have been propounded to explain this behaviour. One of them is the colonial experience, which suggests that as Ghanaians saw the rule of colonial masters as rule by strangers, and they have also applied this notion to any legacy that remains in the country from the colonial masters. It is suggested that this is why, for instance, Ghanaians view stealing of government property as stealing of the property of a stranger. This notion has become deep-rooted among

Ghanaians who, even after independence, have found it difficult to do away with this practice. According to Nana Adjei (Personal Communication, 17 November 2011), this explains why some contemporary Ghanaian politicians continue to steal from the national coffers with impunity–stealing from the national reserves is not viewed as immoral. Others also blame this attitude on the way and manner in which reserves were created. The new system of managing the forests was perceived to undermine customary arrangements for ownership and for the utilisation of natural resources (Dikirr, 2005), but this non-involvement of local people in the management of the forest has encouraged apathy towards what happens to the national assets from people who live at the fringes of the reserves. In fact, these people are mostly the worst offenders in the chain-saw business and in poaching in the forests. This may represent a registeration of their resentment over the effects that the creation of these forests has had on their socio-economic lives. I should point out that it is not only those who live in the fringes of the forest that are the worst offenders but at times people from far away take advantage of the situation.

Yet another reason for the environmental woes of the Berekum Area, according to my informants, is the influence that Christianity has had there. This issue is however discussed in a different chapter (Chapter Six) due to the complex issues that are associated with it. The following section will now deal with the effects of the above environmental problems on the people of the study area in particular, and Ghana more generally.

The effects of the current environmental situation on the people of the study area

During the data collection phase of this research, I observed that the people were already aware of the unacceptable level of environmental degradation in the area, and many found the situation very painful and demoralising. They were emphatic that the degradation that the environment has suffered is gradually bringing about unpredictable weather conditions in the area, and that the area is gradually being taken over by grasses. A prominent foodstuff seller of the main market of Berekum said ‘some time ago, foodstuffs like cocoyam, yam, plantain and cassava used to do very well here, but it is not the case today. At times, we have to travel long distance to bring some [of these] from other areas to meet the local requirement’ (Akua Ampomah, personal communication, 7 October 2011). I also noticed that although my respondents did not have indigenous terms for what have become known today as ‘greenhouse gases’ such as carbon dioxide, methane, nitrous oxide, their comments suggested that they were aware of the existence of such gases as well as their effects on the environment.

A quote that Chris Park made when he was analysing the 'Gaia Hypothesis'[1] here is relevant, as it echoes the implications of the above situation for Berekum people: 'The Earth will probably survive, no matter what humans do to it, but its survival might not include humans' (Park, 2007: 185).

Loss of primary basic forest products for the local people

Ɔkyeame Okra Acheampong told me that the traditional area is experiencing an environmental crisis. The following words from him tell it all:

> There has been a massive destruction of the environment. For instance, big trees, lianas, all kinds of *nhoma* [ropes] such as *sibire*, *nnɔtɔ*, and leaves like *aworomo* for making *kenkey*, and a lot more which we could used in our daily activities are now gone. It is only at some few places that you can find some. Even wild food like *nnwa* [snails]; *mmire*, *kyikyirikyi* and *nsosoa* [kinds of mushroom]; and common wild fruits like *atefu*–which were in abundance here – are all now vanished in this area, all due to environmental degradation, particularly bushfires. The rainfall pattern has become erratic. This has affected food production. You know this place was noted for its big sizes of plantain, cocoyam, and cassava. The sizes of these crops today are nothing near those of some years past. What can save us is to allow the land to fallow for a longer period to regenerate (personal communication, 6 December 2011).

Kwame Amoabeng, a hunter, corroborated *Ɔkyeame's* assertion, adding that: 'now some animals like *nko* [parrots], *nsee* [a kind of strong beaked bird] and *abobonnua* [woodpeckers] are all gone from this area. Even my kids do not know what *asee* [singular] is, for they are no more here' (Personal Communication, 13 December 2012). Corroborating this further, Nana Oppong Taah Senti II, the chief of Abi, said:

---

[1] This theory or hypothesis was first suggested by James Lovelock in 1979. It was based on the idea that the earth should be seen and treated as a living organism. Lovelock named the hypothesis after the Greek goddess, Gaea-taking from the Greek myth that sees the earth as a person (personifies the earth). The theory assumes that both the living and the non-living part of the earth are mutually interdependent and must be seen as such (see Park, 2007:184).

> There has been a vast change. The sun now shines in the extreme. The wind blows in the extreme. The rivers did not get dried up some time ago, but they get dried today. Now herbs and trees, which were not seen here before are springing up. A lot of animals, such as *esono, nwam, nsoromoa, ekoo, weataa, aperaa,* etc. are no more found here. We are indeed in a deep crisis (Personal Communication, 16 November 2011).

The effects of environmental degradation have manifested themselves in many ways in the study area, with poverty being one of its impacts.

In Ghana, whenever areas associated with wealth and general prosperity are mentioned, Berekum is one of the areas that readily come to mind. This is because the level of poverty is low and manageable in this area in recent history. The area had a lot of successful farmers, and in the mid-1970s, *Ɔpanyin* Dabɔne–an illustrious son of the traditional area–was recognised by the government of Ghana, the Supreme Military Council (SMC 1) by having his picture printed on the old ten Cedis (¢10.00) of Ghanaian currency. Because of the wealth associated with the area, when it is said that somebody hails from Berekum, then (anywhere in Ghana) the next comment (even now) is generally '*ei! enneɛ wowɔ sika ɛfiri sɛ wofiri sikakrom*' (literal translation: 'then you have money, for are you from a town of money/wealth').

However, Berekum is now a shadow of its former self, and the level of poverty there is gradually increasing. The main reason for this is the degradation of the fertility of the land, which the local people depend on for their livelihood. As a result, many of the beautiful buildings–which are one of the reasons that Berekum is associated with wealth–are in a dilapidated state. Many of them have not been painted for years, as there is no money to maintain them. The situation is worse in the surrounding villages where many of the buildings are in an even more dilapidated state, and some are in total ruins. I was told during the fieldwork that some families have decided to sign agreements with the building societies that have sprung up in the country in order to save their family houses from collapsing. These building societies rehabilitate the buildings and rent them out to either individual business people or corporate bodies in order to recoup their investment before handing them back to the owners. This seems to be a way for some Berekum residents to maintain their homes, but it means that family members have to relocate until the sponsors recoup their investments, which can take decades. Finding alternative accommodation is not easy in this area due to an astronomical increase in rent advances.

The link between the above situation and the change in the local environment is that the local economy is largely agrarian (see the 1996

District report). Therefore, the level of degradation is correlated with the living standards in the area. Given such relations, De Clerck *et al.* (2006:553) argue that 'Ecologists must focus research on interactions between ecosystems and societies in developing countries, while development experts need to understand that environmental management is a critical component of poverty alleviation strategies'. This obviously has implications for environmental policy-makers, as an alternative source of livelihood is germane to the success of any policy that is intended to conserve the environment, particularly in developing countries. For instance, the national newspaper, *The Ghanaian Times*, reported on 29 August 2012 that some women in the Church of Pentecost at Apori in Oda (located in the Eastern region of Ghana) have begun leaving the church as a protest to preaching against illegal chainsaw activities. These women are apparently those who are marrying chain-saw operators. This is an indication that the issue of survival is one of the major driving forces in the current deforestation and environmental degradation that Ghana, and in particular the Berekum Area is experiencing.

Moreover, *The Daily Graphic,* Ghana's leading newspaper, reported in its 30th April 2013 edition that Mr Samuel Afari Dartey, the Chief Executive Officer of the Forestry Commission, has said that sixty-two forests and wildlife guards have been killed in fatal attacks by illegal chain-saw and galamsey operators [illegal small scale miners] in the last two years. Policies connected with environmental conservation and the alleviation of poverty are complex, with research showing that environmental conservation measures have increased the poverty levels of local people in some areas due to the limits that these measures place on their access to natural resources (Brockington and Schmidt-Soltau, 2005). This explains why proactive research is needed to bail countries out of their present predicament.

The outbreak of diseases

Environmental degradation has had a serious and negative impact on healthcare delivery in Berekum. The destruction of vegetation means the destruction of medicinal herbs that play a significant role in healthcare delivery (Brown, 1992). Although the traditional area can boast of one of the finest hospitals in Ghana–the Holy Family Hospital (owned by the Roman Catholic Church)–as well as a couple of privately-owned clinics, many in the area are unable to take advantage of modern healthcare provision because of poverty, and rely instead on the use of herbs. This makes the destruction of the vegetation very serious. The high level of pollution in the area is also worsening the health of the people. Most serious is that resulting from the smoke from charcoal producers and the dust from table-saw-mills set up within residential areas.

Many Berekum residents may be unaware of the damage being done to their health by the charcoal producers and the table-saw-milling operating sites dotted all over the communities and the indoor biomass. But research has shown links between premature deaths in the area and the pollutants mentioned above. For instance, Ezzati's (2005) research on 'indoor air pollution and health in developing countries' established that 'emissions released from indoor biomass burning are responsible for 1.6 million premature deaths a year due to acute respiratory infection, pneumonia and other respiratory illnesses' (Ezzati, 2005, quoted in DeClerck, 2006, p. 536). This provides a pointer for the decision-makers in the area regarding how to address the mounting levels of air pollution in the municipality.

In an interview, Mr. Oliver (2011), the Berekum Municipal Environmental Officer, informed me that there is a ban on the mass-production of charcoal in the municipality, but that the enforcement of this has been lax. This laxity may, in part, be due to the astronomical increase in the price of Liquefied Petroleum Gas (LPG) that has recently occurred in Ghana. About two decades ago, it was government policy to encourage the domestic use of LPG, and gas cylinders were given free of charge to those who purchased gas stoves to promote this and serve as an incentive to encourage people to stop using charcoal and other sources of bio-fuel. However, the current high price of LPG has become a serious disincentive for people to use it now, which may explain why law enforcement agencies are not being strict on charcoal producers. The production of charcoal results in a dual damage to humanity–apart from deforestation, the smoke resulting from the burning of the wood for the charcoal is damaging to human health, as Ezzati (2005) has shown.

The emerging field of 'conservation medicine'–which focuses on the intersection of the environment, the ecology of human and non-human hosts, pathogens, and conservation biology–aims to reduce the spread of emerging infectious diseases (De Clerck, 2006: 536). To have a better idea about pollutant-related diseases in the study area, I consultved the Holy Family Hospital–the main modern healthcare provider in the area–and was provided with the following information on pollutant-related diseases from 2007-2012.

Table 4. Pollution-Related Diseases over the past Five Years (2007-2011) in Berekum Municipality (seen in the Holy Family Hospital, Berekum)

Disease	2007	2008	2009	2010	2011
Asthma	164	265	286	296	252
Cholera	1	3	2	6	0
Diarrhoea	234	283	258	247	256
Dysentery	131	179	163	153	157
Typhoid	55	154	275	314	200

Source: Holy Family Hospital Records, Berekum, 2012.

The above table, provided by the main health centre in the study area, summarises the effects of pollution resulting from poor environmental management in the study area.

Natural disasters

Another serious effect of deforestation in Berekum is houses losing their roofs during heavy storms–a phenomenon which, according to my informants, is very rare. However, I was told that on the eve of Easter 2012, a heavy storm swept through Berekum and ripped the roofs off several buildings. This necessitated an urgent appeal to the government through the National Disaster Management Organisation (NADMO), which obviously adds to the strain on national budget (that is already under intense pressure). Nowadays, any serious storm results in turbidity, erosion, and floods in the area, something that is new to the traditional area according to Nana Okofo Dartey (Personal Communication, 14 December 2011).

It is important to state that though anthropogenic activities are a cause of concern when it comes to environment-related problems in Berekum, the general natural changing patterns in climatic conditions across the world cannot go without mention. Having explored the environmental problems, their causes and effects in the current chapter, the next chapter will examine Berekum people's worldviews, which impact on their ecological practices.

Summary

This chapter has discussed Berekum's environmental degradation with regards to the misuse of natural resources and particularly the encroachment on the two national forest reserves in the area. The specific environmental problems in the area were found to be those of deforestation, poaching, bushfires, threats to rivers, and poor sanitary conditions. The discussion also touched on the other major reasons for the current environmental problems in the area. These were the over-reliance on modern (scientific) methods for conserving nature, over-population, and a decline in the authority of chiefs (who are the custodians of natural resources in their respective areas) (see also Mawere 2014). Other reasons identified include a lack of patriotism, general apathy, and the influence of foreign religions such as Christianity within the area. The environmental degradation was found to have resulted in the loss of primary forest products for the local community, an increase in poverty, an outbreak of diseases, and natural disasters (such as buildings losing their roofs without the shelter of trees during storms).

CHAPTER FOUR

BEREKUM PEOPLE'S TRADITIONAL WORLDVIEW

Introduction

This chapter examines the Berekum people's worldview. One specific issue which will be discussed in detail is the religious life of the people prior to their encounter with the Western European religion and culture, as well as the current religious landscape of the study area, which has implications for local ecological management.

The Berekum Worldview

I begin this chapter by examining the worldview of the people under study, as my observations revealed that their worldview forms the basis of their conduct and their conception of environmental conservation. By 'the Berekum worldview', I am referring to the set of assumptions that the Berekum people hold about the universe, which informs their understanding and interpretation of the nature and scope of reality. In other words, their worldview concerns how they perceive or understand the world and their immediate environment which they share with other beings.

Explaining what a worldview entails, Ikenga-Metuh (1985) says that it is 'the complex of a people's belief about the origin, structure and organization of the universe, and the laws governing the interaction of beings in it' (1985: 37), while Elkin (1938, cited in Rose, James and Watson, 2003: 59) sees a worldview as 'a view of nature and life, of the universe and man, which unites them with nature's activities and species' in a bond 'of mutual life-giving'. In other words, a worldview refers to the set of basic assumptions that a people have developed for explaining reality and their place and purpose in this world (Mkhize, 2004). For Bauman et al. (2011), a worldview is 'a set of basic assumptions through which one views reality, usually shared among members of a community and often heavily influenced by religion' (Bauman *et al.*, 2011: 235). Kraft (1989) defines a worldview as the set of 'culturally structured assumptions, values and commitments/allegiances underlying a people's perception of reality and their responses to those perceptions' (1989: 20), adding that worldview is not separate from culture. This affirms the Berekum people's view that worldview is a legacy of the progenitors of their society.

Another word that scholars sometimes use for worldview is 'cosmovision' (Miller, 1999; Slikkerveer, 1999). Slikkerveer for instance, says cosmovision:

> Refers specifically to the way in which the members of a particular culture perceive their world, cosmos or universe. It represents the view of the world as a living being, its totality including not only natural elements such as plants, animals and humans, but also spiritual elements such as spirits, ancestors and future generations. In this view, nature does not belong to humans, but humans to nature. As the concept of cosmovision includes the relationship between humans, nature and the spiritual world, it describes the principles, roles and processes of the forces of nature, often intertwined with local belief systems (Slikkerveer, 1999: 171).

Hence, a cosmovision or a worldview can be described as a mental lens through which people perceive the world (Olsen, Lodwick and Dunlap, 1992, cited in Hart 2010: 2).

According to Nana Kwabena Wusu (Personal Communication, 12 October 2011), the worldview of the people of Berekum traditional society is the sum of their core ideas about the universe and their role within it. This understanding reflects Chima's (1999: 35) view (as cited in Kamwaria and Katola, 2012: 50) that 'World-view is humanity's idea of the universe; the organization of ideas that answers questions such as: Who am I? Among what do I move? What are my relations to other people, things, or events?' Nana Wusu explained further that their worldview is the product of systematic reflections on phenomena and experiences that their forebears have bequeathed to them. Thus, a logical deduction from the above conception of worldview is that it is something that is abstractly constructed, but rather than remaining as something that exists purely at the abstract or conceptual level, it is manifested in real life through informing choices and actions via a people's culture (see Kraft, 1996). In other words, culture acts as a vehicle for the transmission of a worldview. It also follows that any new understanding of a phenomenon or experience will affect the expression or the manifestation of a worldview that is expressed in the culture.

For the people of Berekum Traditional Society, the world is made up of both simple and complex phenomena, some of which are visible while others are not. In order for humans to be able to create meaning in their lives, or to live meaningful lives, they have to be able to understand the various elements in the universe and how they relate with each. Berekum people also conceive of the human condition to be full of uncertainties and insecurities. These insecurities have led humans to realise that they are not 'whole', and thus need to 'lean on' something that can help to give them security. This succour, according to my informants, can be found in the maintenance of a harmonious relationship with the spiritual world. This explains why the spiritual usually influences their life and thought. *Ɔkyeame* Okra Acheampong explained in an interview that:

> We, traditional Africans do not act in a vacuum. There is a reason for every single act that we undertake. We believe that there are two interconnected realities of this world. These are the 'seen' (physical) and the unseen (spiritual) dimensions of life. These realities must necessarily be in harmony for things to move on well. Those who know and understand this reality succeed in their endeavours. It was this understanding that made *nananom* (ancestors) to always connect the 'seen' to the 'unseen' in all their endeavours. Anyone who loses this basic principle will never understand the African (Personal Communication, 7 November 2011).

This statement suggests that the supernatural (or the spiritual) plays a major role in the life and thought of the people of Berekum Traditional Society. Smith's (1986) study on the Akan confirms this view, and Mbiti's research (1969) among other African peoples revealed that they conceive of themselves as having two-dimensional lives, namely, their life in the world of spirits and their life in the world of humans (or the physical world). However, the division of the universe into the spiritual and the physical can be said to be for purposes of convenience, as the two are, somewhat paradoxically, only viewed as one entity. This is because the Berekum people believe that the two worlds cannot exist independently of each other (see Awuah-Nyamekye, 2009b). In other words, the worlds of humans and spirits are not independent worlds, for one has no meaning without the other (Dickson and Ellingworth, 1969). Olupona corroborates this thus:

> African spiritual experience is one in which the "divine" or the "sacred" realm interpenetrates into the daily experience of human person so much that religion, culture, and society are imperatively interrelated. The significance of this interaction is that there is no clear-cut distinction between religious and secular spheres or perspective of the ordinary life experience (Olupona, 2000: xvii).

However, although the spiritual and the physical are indispensable to each other, the world of spirits presides over the world of humans. In the words of Nana Kwabena Wusu (Personal Communication, 12 October 2011), '*Eyi ne sɛ nea Berekumfoɔ hunu wiase fa'*, which translates 'this is how the traditional Berekum people see or perceive the universe and all that is in it to be'. This statement summarises the traditional Berekum people's worldview. The above discussion reveals that the supernatural (religion) plays a major role in the Berekum people's worldview.

Nana Kwabena Wusu further observed that because the Berekum worldview is underpinned by its indigenous religion, entities such as land, water, animals and plants are not just materials of production, but also have a place in their religion or spirituality (PersonalCcommunication, 12 October 2011, at Nana Takyiwaa's palace at Senase). However, *Ɔpanin* Kyere Kwame (Personal Communication, 2011) also pointed out that whilst respect is accorded all elements of creation in the Berekum worldview, those that constitute the *primary* life-sustaining elements of creation are accorded more respect than others. For instance, trees, animals and rivers are fundamental or primary sustaining elements for human life and, for that reason, more attention is given to such objects in practice. Corroborating the above view, Nana Okofo Dartey, the *Gyasehene* of Berekum Traditional Area, observed: 'Humans have a duty to conserve God's creation. That is why one of the duties of a chief is to ensure that the forests in his area are not depleted anyhow. This also explains why we at times earmark a place as a sacred grove' (personal communication, 14 December 2011). Millar (1999) and Gonese (1999) have made similar observations. Giving credence to this view of traditional Africans, Mbiti (1991) says that 'living in harmony with nature is so dear to indigenous African peoples that they tend to assign biological life even to inanimate objects' (Mbiti, 1991: 44). In the African worldview, the universe is the creation of God, and thus, is itself viewed in religious terms. African peoples strongly believe that they have a duty to protect and conserve the universe for both secular and religious reasons. In addition, Mbiti points out that:

> African religion sees nature as a friend of man (humans) and vice versa. He (humanity) is an integral part of nature and the priest of nature. The destruction or pollution of nature (including air, water, forest, land, animals, trees, plants and useful insects) brings harm to all life in general and injuries to human welfare in particular. Therefore, man [humanity] has to preserve nature and use it wisely, indeed mercifully, for his own and its survival (Mbiti, 1991: 44).

The worldview of the traditional African peoples is markedly distinct from the dominant Western worldview (Walker, 2004; Little Bear, 2000; Pichette *et al.*, 1999). Traditional Berekum Society, and indeed many traditional African societies, do not make the clear distinction between what we now refer to as 'religion' and 'spirituality' in Western and North American societies. Numerous scholars now distinguish between the terms 'religion' and 'spirituality', and many people prefer to be identified as 'spiritual' rather than 'religious' (Taylor, 2005). For instance, the sociologist of religion, Wade Clark Roof, in one of his seminal studies, observed that for many, '...to be religious conveys an institutional connotation [while] to be spiritual..is more personal and empowering and has to do with the deepest motivations in life' (Roof, 1993: 76-7, cited in Taylor, 2005). In a study aimed at discovering people's views about the two terms, Helminiak (1996) discovered that 90% of his informants preferred to refer to themselves as spiritual rather than religious, and reported that according to these people, religion:

> Implies a social and political organization with structures, rules, officials, [and] dues [while] spirituality refers only to the sense of the transcendent, which organized religions carry and are supposed to foster (Helminiak, 1996: 33).

Zinnbauer *et al.* (1997) also found in their study that 'religiousness is increasingly characterized as "narrow and institutional," and spirituality [...] as "personal and subjective"' (Zinnbauer *et al.*, 1997: 563). For Taylor (2005), probably the most commonly understood difference between the two terms is seeing religion as "organized" and "institutional", and spirituality as involving one's deepest moral values and most profound life experiences (Taylor, 2005: ix).

As I have indicated above, this distinction does not arise in Berekum Traditional Society as these terms are western constructs, and hence there are no local (Akan) terms for them. Even among some of the Christians I interviewed, the distinction between the terms "religion" and "spirituality" was difficult to make. This may be due to the fact they continue to be influenced by their indigenous worldview. In the view of *Ɔpanin* Kyere Kwame, 'it will be a contradiction to say that "I am spiritual but not religious", for religiosity implies spirituality and vice versa in our tradition'. In the life and thought of the people of Berekum traditional society and for that matter traditional Africans in general, life is seen in a holistic way, with the spiritual and the secular intrinsically fused together. Nel (2008) makes a similar observation. Another scholar that shares the traditional Berekum view is Seyyed Hossein Nasr, an important Islamic theologian and philosopher. In a paper entitled 'The Spirituality and Religious Dimension of the Environmental Crisis' (in Chittick, 2007), Nasr rejects the distinction between the terms "spirituality" and "religion", contending that the term "spiritual" or "spirituality" as it is understood today did not exist until around the mid-nineteenth century, when it began to be applied in its current usage by French Catholic theologians, and this usage later crept into the English Language. He emphasises that today, the term:

> Denotes for many people precisely those elements of religion which have been forgotten in the West and which therefore have come to be identified wrongly with spirituality as distinct from religion. From my point of view, which is always of course a traditional one, there is no spirituality without religion (Chittick, 2007, p. 29).

This clarification can aid an appreciation of the traditional Berekum people's way of life. One should not expect to find all the trappings of Western religion and culture in non-Western societies. Such interpretations lead to misrepresentations, and such misrepresentations or misunderstandings of the African led many of the early missionaries who came to evangelise in Africa to conclude that the African has no religion. It is true that Western education and Christianity have made phenomenal inroads in the study area, and indeed many parts of Africa today, but my observations show that many of the people's lives continue to be influenced by their indigenous religious worldviews as well. Ellis and ter Haar's view that 'religious worldviews do not necessarily diminish with formal education' (2004, p. 51) is clearly seen in relation to the traditional Berekum society, with even the 'highly' educated people in the study area continuing to explain the causes of several events (such as chronic diseases) in traditional religious terms.

I have already established that worldview forms a basis for the conduct of people in the traditional Berekum society. However, there are a number of aspects of worldviews that one should not overlook. One is their susceptibility to change (Hart, 2010), and another is that a sub-worldview often exists under a dominant one (Hart, 2010). This explains the slight variations in religio-cultural practices among indigenous peoples. Factors such as new experiences, increased knowledge and contact with other cultures can bring about changes in a people's worldview. It has also been noted that any new understanding of a phenomenon or experience will affect the expression or the manifestation of a worldview. This means that the traditional Berekum worldview of the 18th century would not have been exactly the same as it is in the 21st century.

The influence of Western education for instance, has created a sub-worldview in the area. The strong presence of foreign religions, such as Islam and Christianity (which is heavily underpinned by Western worldview) has affected the worldview of many people in the study area, and this has implications for ecological management there. This is because issues regarding ecology in Berekum Traditional Society are influenced by the religio-cultural worldviews of the people, and therefore, anything that affects the people's worldview has ramifications for the management of the local ecosystem. This issue is given more attention in Chapter Seven, which deals with the impact of colonialism and Christianity on the indigenous Berekum lifestyles and environmental management.

As a result of the above factors, it may be thought that it is difficult to talk of a single Berekum traditional worldview today. Nonetheless, there is evidence of several continuities existing together with these discontinuities and changes. In other words, one can still make a case for the traditional Berekum worldview and, despite the influence of these factors, the core combination of elements that comprise the traditional Berekum worldview remain intact and continue to direct conduct. These include the belief in a supreme being, the belief in divinities/gods, the belief in spirit beings, and the belief in impersonal (mystical) power(s) that manifest themselves in the workings of magic, sorcery and witchcraft. There is still a strong belief that the spiritual world can influence the life of human beings either positively or negatively, and there is still a strong sense of community solidarity. The traditions for getting married and naming children remain largely the same. The European system of marriage, in which marriage is a legal contract, is practised in Berekum today, but no court in the Berekum Society or Ghana as a whole will register any marriage that has not gone through the customary ceremony. The system of inheritance also remains largely unchanged, as does the procedure for acquiring land.

The resilience of the traditional worldview in Berekum lies in the fact that its people have an attitude that is captured by the saying '*amanmre yento ntwene*', which literally mean 'we cannot throw away our customs and traditions'. Thus, one may argue that the combination of traditional religious beliefs and practices with Christianity in traditional Berekum society (and other traditional African societies) can be traced to the hold that traditional worldviews exert on these people. This is reflected in Laurenti Magesa's (1997) observation that most of the time, African Christians 'seek comfort in their own religious symbol systems, even though these may not correspond exactly to those inculcated and expected by their Christian leaders. Indeed, these are often symbols and rituals that church leaders have explicitly condemned' (Magesa, 1997: 7). Mbiti and

Burleson (1986: 12) put it thus: 'Africans come out of African religion but they don't take off their traditional religiosity. They come as they are. They come as people whose worldview is shaped according to African religion'. As the preceding analysis has shown, Berekum people have a religious worldview, just as Tucker and Grim's (2009) study argues that a culture or people's worldviews are contained in religious cosmologies and expressed through rituals and symbols, which make it imperative to throw more light on some aspects of the religion of the Berekum people. Their religion falls under what is known as African Traditional Religion (ATR). I have already argued that as one of the Bono sub-groups within the larger Akan ethnic group, the Berekum people's religious beliefs and practices have many similarities with the Akan people of West Africa. Traditional Berekum people, as I pointed out earlier, make little distinction between their secular and spiritual lives. Their religiosity is seen in their political, social, economic, ethical, medical and environmental lives. Just as it is in other traditional Akan societies–and even among other traditional Africans–religion is life and life is religion (Opoku, 1978, Olupona, 2000). Gyekye describes the religiosity of the African thus: 'To be born into African society is to be born into a culture that is intensely and pervasively religious and that means, and requires, participating in the religious beliefs and rituals of the community'(1996: 4)

Supporting the view that religious influence runs deep in the life of the people of Berekum, Akua Amponsah (Personal Communication, 16 October 2011, Berekum) noted that 'the knowledge of *Onyame/Onyankopɔn* (God) and His influence on us is taken for granted. That is why we have a proverb that runs '*Obi nkyerɛ abɔfra nyame,*' to wit 'no one teaches the child about God'. This implies that the consciousness of God is inborn, and suggests that the belief in *Onyame/Onyankopɔn* is an essential element in both Berekum and Akan religious thought. A basic belief among the traditional Berekum people is that God is everywhere, hence the common Berekum proverb *'wopɛ sɛ woka asɛm kyerɛ Onyame a, ka kyerɛ mframa* ('If you want to tell God something, tell it to the wind'). One of the attributes of *Onyame* in the Berekum society is *brekyirehunuade* ('He who can know things that are behind or hidden from him'–i.e. omniscience). In the words of *Ɔkɔmfoɔ* Agyemang Serwaa, 'we credit *Onyame* with creation, including human beings, animals and trees. That is why we give names such as *Ɔboadee* [creator], *Bɔrebɔre* [Excavator, Hewer, Carver], and *Amaosu* [giver of rain] to him'[1]. This claim is supported by Ekem (2008), who says that, 'In the Akan spiritual universe, which hardly lends

[1] These words were also used by Danquah (1968) and Opoku (1978: 15).

itself to watertight compartmentalization, *Onyame* is generally acknowledged as the creator under whom created entities carry out their functions in human societies' (Ekem, 2008: 35). *Ɔkɔmfoɔ* Agyemang continues, 'because we believe that everyone was created by God we say *'yɛn nyinaa yɛ Onyame mma, obiara nnyɛ asaase ba'*–'all of us are the children of God no one is the child of the earth' (Personal Communication, 15 December 2011). Corroborating *Ɔkɔmfoɔ* Agyemang's claim, Kwaku Asamoah (Personal Communication, 8 Nov 2011) adds that 'the dignity that traditional believers accord every human being is informed by their belief that everybody was created by *Onyame*', claiming that it is because they believe that everything was created by *Onyame* that they say '*Asansa se nea Onyame ayɛ nyinaa yɛ*'–'The hawk says all that God has created is good'. Asamoah stressed that this view dictates their attitudes towards certain elements of creation, such that they know *Onyame* to be someone who hates stubbornness, and explains why *Onyame* withdrew from human beings after creating the world. There is a myth that runs:

> In the beginning God was very close to humans on earth but the women persistently hit the sky, the dwelling place of God when they pounded *fufu* (an Akan local dish). God persistently complained about this bad behaviour of the women but they were adamant. This annoyed God and thus he withdrew from humans and went further into the sky.[2]

Since his withdrawal, God has entrusted the rule of this world to the lesser deities, which explains why traditional believers consider the *abosom* to be the representatives or ministers of God in his theocratic government, and also why 'we [Berekums] worship these gods, for it is a way by which we honour *Onyame*, our creator' (Kwaku Asamoah, Personal Communication, 8 November 2011).

The Traditional Berekum People's belief in God can be argued to have ecological implications as they take their belief that everything was created by *Onyame* to give them responsibilities for protecting and conserving the environment. Nana Akumfi Ameyaw summarises the influence that the belief in *Onyame* has on the people's attitude towards creation as follows:

> In God's creation there are things that can be used [destroyed] and other that cannot be used, so the destruction of such things constitutes a sin. Even those that can be used have the right time for their use. Anything short of this is considered as a sin in our tradition. This is to ensure that we

[2] This is a commonplace myth in the study area.

> make sustainable use of natural resources (Personal Communication, 7 November 2011).

Studies among other indigenous peoples support this connection. For instance, Hussain (2002) finds that the 'conservation ethics of local communities are complex and deeply embedded in their local cultural and historical cosmologies, and are therefore often indiscernible to, or misunderstood by, outsiders'. Zahan (200: 6) confirms this belief among other West African societies in his essay ' Some reflections on African spirituality' in the Olupona's (2000) edited volume--*African spirituality: Forms, meanings and experiences.*

The Ancestors (*Nananom Nsamanfoɔ*) are another central element in traditional Berekum religion. The ancestors are described as the 'living dead'–the departed members of the community. An interviewee noted, 'We describe them as "living dead" because, although they are dead, we still consider them as an important part of our society, for they continue to influence our lives' (Nana Adwoa Takyiwaa, Kurontihemaa, Personal Communication, 12 October 2011, Berekum Traditional Area).

Nana Kwabena Wusu, the *Twafoɔhene* of the *Kurontihene* of Berekum Traditional Area added:

> It is not every departed member of the family that we confer ancestorship on. To qualify for it, one must have led an exemplary live, married and had children, died a natural death [not through an accident or any violent means] and must have advanced in years among others. Since the ancestors constitute the forebears of the living, thereby being initiators of the core values, rules and regulations of the community–including the traditional ecological knowledge that is in use today–there is no way they [ancestors] will not be mentioned in discussion of traditional ecological knowledge and practices. Herein lies the ecological importance of the belief in ancestors (Personal Communication, 12 October 2011).

Another belief that is central to Berekum people's religion is that of the a*bosom nnuasa* (myriad of lesser deities), which are the multiplicity of other spirit beings in the universe. These spirit beings are believed to inhabit natural objects such as rocks, mountains, caves, trees, rivers, animal. In addition, as Yaw *Kɔmfoɔ* hinted, all the deities in Berekum have particular animals or trees as taboos, and some have more than one animal or tree taboo (Personal Communication, 13 November 2011).3 These beliefs influence Berekum people's attitudes towards natural objects, and

[3] Falconer (1992) also discusses these beliefs in his study among the Asante.

the relationship between natural object and deities explains why contravening any of the taboos in relation to natural objects is a matter of concern to the entire community. This is because the breaking of a taboo is viewed as *mmusuo* (a serious sin), as it can arouse supernatural anger towards the people (Nana Kwabena Wusu, Personal Communication, 12 October 2011, at Nana Takyiwa's palace in Senase). This belief may be explained through the principle of acting for the collective good (collective responsibility), and could be argued to have sustained the endurance of traditional ecological laws in the past as, according to Nana Yeboah (Personal Communication, October 2011), there is a strong belief in Berekum Traditional Society that an individual's acts of omission and commission can both result in collective suffering. For instance, the defilement of the ancestors, local gods or spirit beings can invoke the wrath of the spiritual entity involved, not only on the perpetrator(s), but on the entire community as well. In consequence, members of the society police each other's actions in order to safeguard the overall weal of the community. Such beliefs also predispose each and every member of the traditional society to vigilance. However, Nana Yeboah noted that these beliefs are no longer as strong and widespread in contemporary Berekum as they were in the past because of colonisation and the influence of other religions.

Furthermore, Berekum people's belief in *mmoatia* also has ecological implications. The people in the study area loosely translate *mmoatia* as 'dwarfs', but interaction with those who claim to be in contact with them suggests that they refer to a type of little spirit or fairy of the forest. The origins of *mmoatia* within the traditional Berekum religious belief system are unclear. The word '*mmoatia*' is a combination of two Akan words, *mmoa* (animals) and *tia* (small or short). Thus, *mmoatia* literally means 'short animals'. However, they are generally seen as something approximating a tiny human being, which makes it difficult to stipulate precisely what sort of beings the *mmoatia* are believed to be–animals or esque entities. However, the existence of *mmoatia*, and their ability to influence the behaviour of human beings both negatively and positively is seen as a reality among the Berekum people and the Akan more generally.

An interview with Kofi Mensah, who is known as *mmoatiakɔmfoɔ* (the priest of dwarfs) in his community, Koraso, threw some light on the belief and activities of *mmoatia* in the area. He said that he has been in league with *mmoatia* for the past 37 years, and reports that although he has never physically seen them before, he hears them when they speak through whistling. He claimed to have first come into contact with *mmoatia* during one of his hunting expeditions. While in the forest, he saw a *konko* (a small ant house), which he destroyed, and then, heard a voice saying '*yɛbɛfa wo'*, meaning 'we will take you'. He could not see anybody around, and on returning home, he said that he saw visions and heard voices in his sleep commanding him to do all sort of things. From that time onwards, he reported receiving accoutrements or paraphernalia on a daily basis from the *mmoatia*. Most of these were shown to me at his shrine. He said that he had been a Methodist before his encounter with the *mmoatia*, and that he had attempted to sever his relationship with them so that he could return to the Methodist church, which he did by throwing away all the accoutrements they had given him. But, in his own words, 'I ended up in a hospital. I nearly died. So I quickly went back to them'. He told me that his association with the *mmoatia* has made him believe that they are not by nature evil, as he was made to believe prior to his encounter with them. He said that they hate wanton destruction of nature, in particular forests and animals such as *koterɛ* (ground agama), *pompoyini* (red-headed rock agama) and *ntatia* (ants).

My interlocutors also claimed that forests and mountainous areas are very important to the *mmoatia*, as they serve as their abode, and that the *mmoatia* would sometimes ask him to carry out sacrifices for rivers. He stated that the *mmoatia* are a major source of traditional medicine and help those who respect them, but only deal with evil-minded people in the community. He added that most of the *tumi* (power) in the *abosom* are from the *mmoatia*, which makes the *abosom* inferior to the *mmoatia* in Akan religious thought (Personal Communication, 16 December 2001, at his shrine at Koraso). This also means that *mmoatia* play a key role in *abisaa* (divination)– a key means of discovering the unknown in the study area (see Awuah-Nyamekye, 2012b).

A critical analysis of the above narrative shows that *mmoatia* are conceived of as eco-centric beings–i.e. they have a central interest in the preservation of nature. The destruction of even the *konko* (the ant house), and the subsequent possessing of Kofi Mensah by them attest to this. The fact that forests are associated with *mmoatia* means that the belief in *mmoatia* has ecological implications for members of the community, and this is also seen in the fact that the origins of most sacred groves in the study area are linked with *mmoatia*. This link makes people afraid to enter such forests, and this prevents them from tampering with the environment in degrading manner.

An important feature of Berekum religion–which is shared by other African peoples (Smith, 1950; Parrinder, 1961)–is the hierarchical arrangement of spiritual entities in their religion. This is headed by the *Onyame/Onyankopɔn* (who is also known by other names), with the next in rank being the *Nananom nsamanfoɔ* (Ancestors), then the *abosom nnuasa* (the lesser gods/deities). The base of the hierarchy is occupied by *asunsum* (impersonal forces), which manifest themselves in the workings of *nnusɛm* (broadly translated as magic), *abayisɛm* (witchcraft), and *adutosɛm* (broadly translated as sorcery), *mmoatia*, and *nsumansɛm* (charms and amulets–fetishes) (*Nyina* Kofi *Bosomfoɔ* Kwame Agyemang, personal communication, 23 November 2011, at his residence at Koraso). This information was corroborated by other *akɔmfoɔ* (traditional priests) (see also Williamson, 1965: 104).

One contested issue here concerns the position of the *mmoatia* and the *abosom*. According to Kofi Mensah–*mmoatiakɔmfoɔ*–the *abosom* are inferior to *mmoatia* as they receive their *tumi* (power) from the *mmoatia*. This, however, contradicts traditional Berekum and Akan beliefs. As was noted earlier, the traditional Berekum people believe that the *tumi* of the *abosom* is derived directly from *Onyame*. This explains why they view the *abosom* as the ministers in God's theocratic government, which should place them above the *mmoatia*. I tried to cross-check the veracity of *mmoatiakɔmfoɔ's* assertion, but nobody else corroborated it. A common view, however, is that the *abosom* are in league with the *mmoatia*. I, therefore, consider *mmoatiakɔmfoɔ*'s assertion to be a result of the aggrandisement of the *tumi* of *mmoatia*, probably as a result of his association with them, given that his assertion is not confirmed by traditional Berekum religious thought.

The hierarchical order of the spiritual entities in both the traditional Berekum and traditional Akan worldviews implies that the cultic attention they give to these supernatural entities varies in degree. That is, the higher the position of the entity, the higher the cultic attention it receives. They also believe that there are both good and evil spirits in the pantheon of spirit beings but all these entities qualify for cultic attention because they are capable of influencing the life of humans either positively or negatively. In the words of Kalu (2000):

> Human beings manoeuvre to tap the resources of the benevolent spirits to ward off the machinations of the devouring spirits. To achieve this vicarious goal, human beings weave enduring covenants with these spirits (Kalu, 2000: 56).

From my focus group discussions with traditionalists (i.e. adherents to the traditional religion), I realised that the heavy reliance on the *abosom* in matters of the administration of their society stems from the close link that they believe to exist between the *abosom* and *Onyame*. What I gathered from the interviews made me believe that it may not be far from accurate to describe the system of government among the traditional Berekum people as the 'rule of the *abosom*'. The chiefs, clans and family heads rely on the advice of the *abosom* through the oracular consultations they make when major decisions are to be taken in their communities, even today. For instance, during the selection of a chief, the *abosom* are consulted through divination in order to ascertain which of the contestants is preferred by the ancestors. Even after the chief is selected, he or she has to rely on the advice of the *abosom* for the day-to-day running of the community (Awuah-Nyamekye, 2009c).

My informants also emphasised that the gods and the ancestors sometimes fail them, but that *Onyame* does not. They cited herbalists and priests at the shrines telling their patients '*Onyame boa a wobɛ te apɔ*' ('if God permits, you will get well'), in addition to everyday sayings such as '*Onyame boa me*' ('God help me'), 'w*o ne Onyame nkɔ*' ('go with God') as examples. These sayings suggest that the traditional Akan people conceive God to be supreme, and *Onyame* to be the ultimate recipient of all the sacrifices, prayers and the cultic attention they offer (All the traditional priests interviewed shared this view).

Berekum religious life has an emphasis on being communal rather than individualistic, which is evident in the cultic attention given to *Nananom nsamanfo* (ancestors) and the *abosom*. This cultic attention is seen on sacred days such as *Akwasidae* (Sunday *Adae*) (for details on Akan *Akwasidae,* see McCaskie (2002) and the annual festival days for the gods and ancestors. The annual festivals are occasions for the people of Berekum to demonstrate a renewal of their commitments to their objects of worship. On these occasions, people redeem their pledges or promises to the spirit beings as well as making new ones. The festivals are used to show their gratitude to their gods and their ancestors for providing them with protection over the previous year. They also use these occasions to ask for protection and blessings from their gods and ancestors for the coming year (Nana Okofo Dartey, Personal Communication, 14 December 2011, at his palace in Berekum). Although the dominant form of cultic attention in Berekum is communally-based, it is also very important at the personal level, especially for individuals who have their own personal deities.

Two main types of *abosom* are held to exist in Berekum Traditional Area. The first category comprises *tete abosom* (nature gods) (This category was mentioned by all the traditional priests interviewed) which Opoku Asare refers to as 'ancient tutelar divinities' (1978: 55). They are communally owned, and have been worshipped from time immemorial (*Ɔkɔmfoɔ* Kwame Agyemang, personal communication, 14 October 2011, at his shrine in Jinijini). *Nyina* Kofi *Bosomfoɔ* Kwame Agyemang corroborated this categorisation (Personal Communication, 23 November 2011). This type of *abosom* is believed to be the children of *Onyame*, acting as his representatives on Earth, and are found in every town and village in the traditional area. A god in this category is known as a *kurobosom* (town or village god). Some towns or villages have more than one of them, in which case they are referred to as *kuroabosom* (town or village gods). Among these gods are two (*Afuofu* and *Tankwasi*) that are considered as *ɔman abosom* (*ɔman* [state] gods). That is, these gods belong to the whole traditional area, and festivals for these two gods have to precede the--*Kwafie*–the main festival for the traditional area (Ɔkɔmfoɔ Yaw, Personal Communication, 7 November 2011, at his shrine in Fetentaa). *Afuofu* is housed in the palace of the *Ɔmanhene* (Paramount Chief) of Berekum Traditional Area, whilst *Tankwasi* is housed at Biadan, a village located about three kilometres from Berekum town.

The second category of gods in Berekum is *abusua abosom* (clan or family gods). The influences of these personal gods are limited to the clan, family or the individual respectively. These gods have recent origins, since many of them take the form of *asuman/bansere* (charms and amulets) (Yaw *Kɔmfoɔ*, Personal Communication, 7 November 2011, at his shrine in Fetentaa). The general belief among the traditional Berekum people is that all these types of gods play a role in God's theocratic government, and thus require human attention.

As *Onyame* is credited with creation, the traditional Berekum people believe that there is a bond between human beings and other creatures. This relationship is a spiritual one, because all life within creation is viewed to exist in a unitary form due to the possession of a *sunsum* (spirit) that is suffused within life by *Onyame*, the source of all creation. Awolalu (1972) suggests this relationship marks the idea that human beings are not created to live alone, but in a relationship with other creatures, while Turaki sees human beings and other creatures–especially plants and animals–as having 'their own existence and place in the universe as independent parts of a whole' (Turaki, 1999: 95). This implies that human beings and other creatures must co-exist in a symbiotic way, a view that Graham (2002) refers to as the 'relational worldview'.

According to the Akan myth of creation, *Odomankoma* (another name for God) created the universe as follows: He created the sky, followed by the earth, rivers and plants, and finally human beings and animals. The animals were to feed on the plants and then provide food for human beings. God realised that human beings needed protection in their environment, and thus created the spirits of waters, forests and rocks to safeguard them (see Opoku, 1978). This creation story leads both the Akan and the traditional Berekum to view the universe as a religious entity, and provides the guiding principles about creation for both the traditional Berekum and the Akan people.

Moreover, it strongly impacts on the traditional Berekum people's attitude to the non-human elements of creation, constituting the framework within which their ecological knowledge and practices are analysed in this book.

The Contemporary Berekum Religious Landscape

The previous sections have demonstrated that religion is a central part of the culture of the people of Berekum Traditional Society, and until 1905, African Traditional Religion was the only religion known and practised in traditional Berekum society. But today, due to their contact with the outside world through colonisation, Western education, foreign religion, and other meaning-giving systems such as humanism, the traditional area is religiously and socially plural. This means that many of the beliefs and practices outlined above have been affected, some being held less widely, some dying out completely, and others being modified into hybrid views through being combined with concepts and ideas from different faiths to form new ideas.

The Current Religious Demography of Berekum Traditional Area

During the data collection phase of the study, I observed that Berekum now contains Traditionalist (indigenous), Christian, Muslim and Bahaist religious communities. However, a survey conducted in the study area by the District Assembly in 2002 provides the following data.

Table 5. Berekum Religious Demography

Religion	Percentage (%)
Christianity	95
Islam	2.5
Traditionalist	1.4
Others	0.9

It is important to note that relying on religious demographic figures in contemporary Berekum/Ghanaian society may be inaccurate, if not highly deceptive, as, at face value, almost everybody is either a Christian or a Muslim but, functionally; the majority of the people still retain many of the beliefs and practices of their indigenous religion. As Asamoah-Gyadu (2003, p. 3) observes: 'People accommodate in their personal worlds as many religious persuasions as they can for as long as they respond to their existential concerns'. Ikenga-Metuh (2002) makes a similar remark about religion in Nigeria, suggesting that hybridity is a contemporary feature of the entire African religious landscape. Moreover, the numbers in the above table only add up to 99.8 percent, perhaps neglecting to mention Baha'i, which was present in the study area at the time that the survey was conducted.

During fieldwork, I also observed that some of my respondents were not being truthful about their religious affiliations. For instance, in the focus group discussions and joint interviews, the first person's answer seemed to influence the responses of the other member(s) of the group, especially when the first respondent claimed to be a Christian. This became clear when they began to answer my questions, as I could tell from their responses what their actual religious affiliations were. I frequently asked those informants why they were not truthful about their religious lives, and the common answer was, 'that is the trend now'. This is because Christians abuse them for saying they belong to the *nananom* religion–i.e. African Traditional Religion. But they also pointed out that surprisingly those who claim to be Christians quickly seek help from the traditional priests anytime that they (Christian converts) face misfortunes beyond their control. This confirms that the people have not abandoned their indigenous beliefs in spite of their conversion to a new faith—Christianity. But the issue is, why should they abuse their colleagues for being members of the religion which they themselves (Christian converts) have not repudiated completely and even resort to whenever in trouble? This reaffirms the view that 'Africans come out of African religion but they don't take off their traditional religiosity' (Mbiti and Burleson, 1986: 12). This attitude provides one explanation for why the indigenous religion persists in the area despite the inroads of Christianity. This issue will be revisited in the course of the discussion.

The Traditionalist Community

The main traditional worship centres in the traditional area are found in the various shrines located in the towns and villages. Some of these shrines are not as active as they used to be and some have become dormant where there is a lack of worshippers to assist the priests, while cultic attention has completely ceased in others due to a lack of priests. The following table lists some of the deities and their shrines, as well as their current status.

Table 6 Some deities in the Berekum Traditional Area

No	Name of community	Name of Deity/Shrine	Current status of cultic attention
1	Abi	*Sira*	Dormant
2	Abisaas.e	*Adampa*	Dormant
3	Adom	*Tigare*	Dormant
4	Akroforo	*Afuofu*	Active
5	Amomaso	*ɔheneba Kwadwo*	Dormant
6	Biadan	*i Tankwasi* *ii Tankwabena* *iii Nsuo Yaa* *iv Nkonkonsa*	Dormant Cultic attention has ceased Dormant Dormant
7	Berekum (Central)	*Nana Mase*	Active
8	Botokrom	*Sakati Kwaku*	Active
9	Domfete	*i Bentin* *ii Boabe* *iii Nangro*	Dormant Cultic attention has ceased Cultic attention has ceased
10	Fetentaa	*i Tankwadwo* *ii Tanyaw* *iii Nimo* *iv Akwatia*	Dormant Active Active Dormant
11	Jinijini	*i Tanframa* *ii Boaban*	Active Active
12	Kato	*i Ntoafri* *ii Nana Gyabi*	Cultic attention has ceased-- , shine in ruins Cultic attention has ceased –shine ruins
13	Koraso	*Ngora*	Active
14	Kotaa	*Tandwomo*	Active

15	Kutre No. 1	*Taakora*	Active
16	Mpatapo	*Asubɔnten*	Active
17	Mantukwa	*Ahu*	Dormant
18	Mpatasie	*Kontɔnkori*	Active
19	Namasua	*i Benyade* *ii Boɔ kofi*	Dormant Dormant
20	Nanasuano	*i Nangro* *ii Ntensere*	Cultic attention has ceased Active
21	Nkyenkyemamu	*Sakati Kwaku*	Dormant
22	Nsapor	*ɔheneba*	Dormant
23	Senase	*i Tansie* *ii Kosua Kra*	Active Dormant
24	Tewbaabi	*Awasu*	Cultic attention has ceased

Source. Author's fieldwork notes

In addition to the above shrines, I discovered that there are some rivers in the area that still receive cultic attention. These rivers are believed to be deities, and thus, some areas along their courses have been designated as *asɔnoyɛso* (shrines), where the cultic attention that the local people refer to as *sɔno* or *gugu* takes place. Maame Akua Donkor explained to me in an interview that, 'most often, divination takes place at such *asɔnoyɛso*. Also, on *nkyinda* [literally 'sacred days'], the priests or in most cases, the attendants go to the *asɔnoyɛso* to offer libation and sacrifice to the river deity concerned to ask for favours' (Personal Communication, 28 January 2011). According to her, some of these favours include protection and prosperity for members of the community in which the river is located. She added that it was common to see people from both near and far trooping to the *asɔnoyɛso* (shrines) on sacred days to present one petition or the other, or to redeem a vow taken before the river deity. Thus, numerous food sources, such as eggs, fowl, sheep, and goats are seen at the shrines on such days. Maame Akua Donkor referred to *Asuo Koraa* (The River Koraa)–the river deity she serves as an attendant–as the most powerful river deity for helping to retrieve stolen property and for protecting people in the traditional area. She said that *Asuo Koraa* has four separate *asɔnoyɛso* located in the four communities through which it runs–Koraso, Jamdede, Biadan and Senase. Although the terms *abosomfie* and *asɔnoyɛso* are both translated into English as 'shrine', there is a difference between the two terms. In an interview, Yaw *Kɔmfoɔ* (Personal Communication, 25 November 2011) explained that the term *abosomfie* (literally, 'the house of the deity/god') is reserved for the *abosom* (deities/gods) that are housed within the towns and villages–that is, whose shrines are located in the residential areas of the communities–while *asɔnoyɛso* are usually located in the bush, where rituals are performed for the spirit(s) believed to be residing in rivers, trees, mountains or any other natural object believed to be the abode of a spiritual entity. He added that those who serve at the *asɔnoyɛso* are normally referred to as attendants, while those at the *abosomfie* are referred to as *akɔmfoɔ* (priests or priestesses). The other important *Asuo abosom* in the traditional area that I discovered during the study include: *Takyiasua, Aprapono, Biadan, Kyimira, Amoma, Awusu, Awusubire, Buasu, Nsapɔ, Sakati,* and *Gyansakyiwa.* This suggests that river deities, apart from serving the religious and social needs of the people, also have ecological value. This is because no farming or hunting activities are permitted in the areas designated as *asɔnoyɛso.* Even today, fishing remains strictly banned in the Koraa River.

Tree Deity

One important tree deity that still continues to receive cultic attention in the Berekum Traditional Area is *Nyina Kofi*. *Nyina* is the local name for the silk cotton tree (*Ceiba pentandra*). Nana Kwame Agyemang, the *Nyina Kofi Bosomfoɔ* (priest) explained that the spirit that is believed to be dwelling in the silk cotton tree is a male, and that its natal day is a Friday, and thus the name *Kofi* is given to it. Kofi is a name that is given to every male born on Friday in the Akan society of Ghana. Friday is thus its sacred day and, on that day, the priest and other worshippers go to the spirit's shrine under the tree to offer libation and sacrifices. *Nyina Kofi* is located at Koraso, about four kilometres from Berekum town, and receives cultic attention from the people of Koraso. Details on *Nyina Kofi* grove are provided in chapter 5.

Figure 7 Nyina Kofi Bosomfoɔ (priest) getting ready to pour libation at the shrine in the grove

Asaase Yaa (the Earth goddess)

One important belief held by both the people of traditional Berekum society and the Akan in general is that the earth (land) is a goddess, and as with *Nyina Kofi,* Yaa, the name given to an Akan female born on a Thursday, means 'the earth'. That is, the people believe that the natal day of the earth is Thursday, and this is therefore designated as her sacred day. As with all goddesses, the people treat her in a special way. Certain things are considered as taboo(s), or hateful to her. These include tilling the land on her sacred day (Thursday), sexual intercourse in the bush, not offering a libation prayer before a grave is dug to bury a corpse. According to Nana Adwoa Takyiwaa, the *Senasehemaa* (Personal Communication, 12 October 2011), a breach of any of the above acts is considered as defilement, and is seen as a serious sin against the earth goddess. She emphasised that it is through purification rites that the severed relation can be restored, or else misfortunes will follow the perpetrator(s) and, in some cases, all the members of the community may suffer for example (see Awuah-Nyamekye, 2009a), strange diseases or poor crop yields. This was corroborated by Nana Ansu Gyeabour (Personal Communication, 11 November 2011). This shows the anthropomorphism present in the belief system of Berekum people.

Another religio-cultural practice is the protection of sacred groves, although this protection has weakened in recent Berekum history (a detail account of sacred groves is provided in chapter 5). These groves are usually patches of land where ordinary people are not allowed to enter because these groves are considered sacred, even if the entry is for religious purposes. In Berekum, only the *Ɔmanhene* or his accredited representatives and the grove attendant are allowed to enter the *Banim* (royal mausoleum) sacred grove at Pepase, although people with various problems are sometimes also granted permission to enter in order to ask for protection. Yaw *Kɔmfoɔ* informed me (and all my key informants confirmed this) that prior to the Berekum people's encounter with the Western world and its culture, all the towns and villages in the

Figure 8 The researcher in an interview with Yaw Kɔmfoɔ of Nimo shrine at Fetentaa

traditional area had their own *abosom* (deities/gods) and sacred groves. But today, due to a number of influences (including science, technology, foreign religions, Western influence.), some of the *abosom* in the towns and villages are dormant as the result of a lack of priests/priestesses and worshippers (as has been noted earlier), and many of the sacred groves have been reduced in size or destroyed completely (Personal Communication, 19 November 2011). This issue kept recurring as I gathered my data but, paradoxically, the influence of these gods continues to affect the lives of the people in the study area. For instance, although the cultic attention given to the River Koraa has reduced, people in the area will still not eat fish or crabs from the river for fear of the repercussions that such acts may draw from the spirits.

Christianity in Berekum

20th January 1482 has been traditionally acknowledged as the date on which Christianity was first introduced to the Gold Coast (now Ghana). This was the day after a Portuguese expedition of six hundred men under the command of Don Diogo d 'Azambuja landed at Elmina, near Cape Coast in Ghana, and on this day they:

> Suspended the banner of Portugal from the bough of a lofty tree, at the foot of which they erected an altar, and prayed for the conversion of the natives from idolatry and (sic) the perpetual prosperity of the church they intended to erect upon the spot (Groves, 1948: 123).

Figure 9 The Emmanuel Presby Church, Berekum

After its introduction in Ghana, it took over four centuries before Christianity reached Berekum. The records show that the first church was set up in Berekum in 1905, and was a Presbyterian Church. According to the Presbyterian Church records (Emmanuel Presby Church, Berekum: Centenary Celebration and Thanks Giving, 10 September 2006), 17th April 1905 is recognised as the day on which Christianity was officially planted in Berekum through the efforts of *Ɔpanyin* Bediako, Reverend Parregaux and Reverend N. T. C. Clerk. Prior to this, and with the encouragement of

Reverend Parregaux, Reverend Clerk arrived at Berekum (on the 2 September 1904) to see Nana Kwabena Wusu–the then *Ɔmanhene* of Berekum–in order to ask for permission to establish a church in Berekum. There were difficulties at first, but permission was finally granted, and the church was set up in 1905. Incidentally, the Emmanuel Presby Church was the first Presbyterian Church in the whole of the Brong Ahafo region of Ghana.

Figure 10 The Freeman Methodist Church, Berekum

Figure 11 The ST. Theresa Catholic Church, Berekum

The next Christian denomination to follow the Presbyterian was the Wesleyans. Oral history has it that the Berekum Freeman Methodist Church was founded in 1920 (Very Rev. Acheampong, personal communication, 18 January 2012). The source mentioned that *Ɔpanyin* James Kwadwo Tabiri was instrumental in its establishment. According to the church's history, *Ɔpanyin* James Kwadwo Tabiri paid a visit to his brother–one M. C. Baiden, who was a teacher-catechist of the Methodist church at Sewia in the Ashanti Region of Ghana–and became converted. Upon the advice of his brother, he began dawn-preaching on his return. His efforts paid dividends, as he soon had other converts, who began as members of the Wesley Society of Berekum, which is now called the Freeman Methodist Church. The Roman Catholic followed the Methodists in 1923. The beginnings of Catholicism in Berekum were the result of a miraculous healing that *Ɔpanyin* Anthony Kwabena Nsia received through the prayers of a Catholic Priest at Jukwa, in the Central region of Ghana. He and his brother *Ɔpanyin* Kwabena Kuma went to Jukwa as masons, but he reportedly fell sick and was healed by the priest. Upon their return home to Berekum in 1923, they began to practise their new faith, during Nana Barnie's time as the *Ɔmanhene* of Berekum.

The Assemblies of God Church followed the Wesleyan. According to Reverend Adu Yeboah–the current minister in charge of this church–the Berekum branch was established in 1945, and Pastor Emmanuel Appiah was its first minister, assisted by Pastor Okye.

The Church of Pentecost reached Berekum in 1959 according to oral sources. The first assembly was in Kato, but in the same year, the second assembly was also established where Golden City Park now stands in Berekum (Pastor Ntiamoah, Personal Communication, 29 January 2012). In an interview with J. S. Inkabi, it was learnt that the Berekum branch of the Anglican Church was established in 1972 through the efforts of Catechist Asiedu and Elders Oduro and Agyemang.

There are now many Christian churches in the Berekum Traditional Area, but only thirty-two of these are on the official register of the Local Council of Churches (LCC) in the traditional area (Right Reverend Bawa, the Methodist Church Circuit Pastor, Personal Communication, 15 January 2011, Berekum). According to the Very Reverend Father, Augustine Obour of St. Theresa's Parish in Berekum, the mapping project he conducted on churches in Berekum town alone revealed that it had one hundred and twenty-six churches in 2009. The Roman Catholic, the Methodist and the Presbyterian churches now have branches in almost all the 33 villages forming the traditional area, with the Methodists having branches in twenty-two villages. Some of the current churches in Berekum are listed below:

Table 7. Christian Churches in Berekum and their categorisation

No.	NAME OF DENOMINATION	CATEGORY
1	Roman Catholic Church	Catholicism
2	Methodist Church	Protestant
3	Presbyterian Church	Protestant
4	Anglican Church	Protestant
5	Assemblies of God	Protestant
6	The Twelve Apostles` Church	African Independent Church (AIC)
7	Seventh Day Adventist Church	Protestant
8	Mozama-Disco-Christo Church	AIC
9	Christ Apostle Church	Protestant
10	A.M.E. Zion Church	Protestant

11	Church of Pentecost	Penteco-Charismatic
12	Apostolic Church of Ghana	Protestant
13	African Faith Tabernacle	AIC
14	Brotherhood Church	AIC
15	Church of the Lord Brotherhood	AIC
16	House of Faith	Penteco-Charismatic
17	Christ Redeemer Church	Penteco-Charismatic
18	Voice of Deliverance Church	Penteco-Charismatic
19	Intl. Central Gospel Church	Penteco-Charismatic
20	Redemption Church	Penteco-Charismatic
21	Holy Temple Ministry	Penteco-Charismatic
22	All for Christ Ministry	Penteco-Charismatic
23	Global Evangelical Presbyterian	Penteco-Charismatic
24	Light of the Lord Church	Penteco-Charismatic
25	*Osamadi* Church (*Gyedie*)	AIC
26	Jesus Redemption Church	Penteco-Charismatic
27	Spirit Alive Church	Penteco-Charismatic
28	Word Anointing Ministry	Penteco-Charismatic
29	Vision Charismatic Church	Penteco-Charismatic
30	Christian Praise International Church	Penteco-Charismatic
31	Agape Evangelical Ministry	Penteco-Charismatic
32	Vision Charismatic Chapel	Penteco-Charismatic
33	Saviour Church (*Gyedie*)	AIC

Source: Author`s field notes, 2012

Christians relationship with the traditionalists

Missionary Christianity has always seen indigenous African religion as a false religion. Christians in Africa have looked down on African Indigenous Religion viewing its practices and beliefs (in spirits that reside in trees, fields, mountains, streams and other natural objects) as forms of paganism or idolatry. This attitude has sometimes resulted in tensions and violent conflicts between Christians and the indigenous communities in the traditional area. Most of the *akɔmfoɔ* that I interviewed said that Christians practise unprovoked attacks on them by conducting their dawn preaching, which is meant to denigrate believers of indigenous religion close to their (*akɔmfoɔ*) residences. At times, the actions of some of the so-called 'born again' Christians result in unnecessary confrontations with the traditional authorities. For instance, *Ɔkyeame* Okra Acheampong told me that a little over a decade ago, some young men in Biadan who were members of the Seventh Day Pentecostal Church defied a long standing community ban on visiting *bɛduruyɛ* (a section of the Koraa River) on Fridays to fetch water. The angry youth of the town 'arrested' these young men and sent them to the chief's palace, where they were flogged for their actions. This created tensions in the community (Personal Communication, 6 December 2011).

From my interviews with the chiefs, I discovered that the challenges that some Christians were creating for their authority have now subsided. The only Christian group that the chiefs cited as still causing occasional trouble were the Pentecostal group. This group of Christians believe that any compromise with the traditionalists in matters of their (Christian) belief is tantamount to idolatry. It must be noted, however, that none of these problems necessarily suggest that Christians in the study area are unconcerned with the current environmental problems in the area, and a section of Chapter 6 outlines the contributions of Christians in the area towards addressing these problems.

The Muslim Communities

The exact date of Islam's introduction to Ghana is not clear. While Fisher and Trimingham believe that the people of Ancient Ghana were converted to Islam through their conquest by the Almoravids, Clarke and Conrad (among others) think otherwise (cited in Acquah, 2011, p. 129). Despite the uncertainty over when Islam was introduced to modern Ghana it is, however believed to have stemmed from the late 14th and early 15th centuries (Acquah, 2011, p.139) through the activities of the Wangara, Dyula and Yarse Muslim traders from the upper Niger, who came to settle in the northern regions of Ghana. The Gonja, Mamprusi, Dagomba and Wara communities were believed to be those that came under the influence of Islam first in modern Ghana (Acquah, 2011, p. 139).

Introduction of Islam in Berekum

The lack of certainty over the exact date of Islam's introduction to Ghana is mirrored in the Islamisation of Berekum, as there is little information on how Islam was introduced into the area at all. When I contacted the current Chief Imam of Berekum, he directed me to one of his assistants, Mujib Mohammed, who happened to have written a dissertation on Islamic education in Berekum. He told me his attempt to get materials on the advent of Islam in Berekum had proved futile, and that he had instead relied on oral tradition. However, he did provide me with a photocopy of the relevant portions of his undergraduate dissertation. According to Mujib, who relied for his information on *Mallam* Asomah (the proprietor of Usmaniya Islamic School in Berekum), Islam emerged in

Figure 12 Muslim Youth helping in the renovation of the Central Mosque in Berekum

Berekum through the activities of Muslim migrants from the northern parts of Ghana who moved southwards in search of greener pastures and better living conditions. The first group of migrants to settle in the area were the Mossi, who were followed by the Zugu from neighbouring Benin and Togo. Dyula-Wangara and Hausa traders also settled in the area because of the abundance of cola nuts in the area that they could trade in. However, many of them were repatriated to their home countries during the 1969 Aliens Compliance Order that was issued during Dr. K. A. Busia's regime. Another branch of Islam found in the study area is that of the Ahmadiyya Movement, which was introduced to Berekum in 1960 by Amadu Kramo (of Amadu Kramo stores fame), Ibrahim Aboagye and Mohammed Assenso (Personal Communication, 29 January 2012). This means there are two Muslim communities in the Berekum Traditional Area–the Orthodox Muslims and the Ahmadis. The Ahmadis are in the minority, and the orthodox are divided between the *Al-sunna* (Sunni) and the Tijaniyya schools of thought. The Ahmadi community has been stable and peaceful since their arrival in Berekum in 1960, whilst the orthodox community experienced some minor internal conflicts until 1999, when a major conflict broke out.

Traditionalists-Muslim Relationship in Berekum

My interviews revealed that theological relationship between Muslims and traditionalists in Berekum is better than that between traditionalists and Christians. The only area over which Muslims seem to strongly diverge with traditionalist views is with regard to the serious cultic attention that the *abosom* receive from traditionalists. All the traditionalists interviewed were unanimous in reporting that they experience few problems with the Muslims, as they agree with each other on many issues. Yaw *Kɔmfoɔ* (Personal Communication, 13 November 2011), for instance, mentioned that they agree on practices such as polygamy, casting spells on evil doers (especially thieves), magical healing and divination. Lewis (1967) reached similar findings in his study.

Divination is not accepted in the official Islamic scripture, with the Sura 5:90 (Al-Ma'idah), for instance, emphatically rejecting divination of any form within Islam. In addition, there are several Ahadith that speak against divination, one saying: 'Whoever goes to the soothsayer and believes in what he says has denied what was revealed to Muhammad' (al Qaradawi: 239). However, it is common knowledge that some Muslim clerics in Berekum practise divination and use charms and amulets. I learnt from my informants that the marriage charms and amulets prepared by Muslim clerics (popularly known in the area as *mallams*) are very effective. As a result, the majority of the *mallams*' clients in the area are women. The good relationship between the Traditionalists and their Muslim counterparts was confirmed by Nana Okofo Dartey, the *Gyaasehene* of the traditional area. He said the Muslims in the traditional area always respect the taboo of the land, and that there has never been a case in the area where a Muslim has violated any taboo, such as entering a sacred grove or going to the farm on a sacred day of a deity in the area. Lewis (1967) also finds that right from the outset, Islam adopted a more flexible attitude towards African beliefs and practices than Christianity.

The Muslims' positive attitude towards environmental protection in the area was confirmed in an interview with Mujib (12 November 2011). Mujib stressed that Islam has a lot to say on the issue of the environment, and many of these prescriptions and lessons can be found in the Hadith. For instance, he mentioned that Mohammed, the Holy Prophet, advised his companions on the need to protect the environment in which they lived, and encouraged them to plant trees. Mujib mentioned that during the defensive battles, the Holy Prophet also urged his companions to desist from the old practice in which a victor had to destroy everything in the land of the vanquished, including cutting down all the trees in the area. Mujib cited the teak plantation near the Usmaniya Primary School in Berekum as a concrete way of putting the Holy Prophet's advice into practice. These attitudes were also confirmed in the Muslim declaration on nature at Assisi.

The Baha'i Community–A brief history of Baha'i Faith in Berekum

In an interview with Nana Kuma Acheampong, a Bahaist and also the *Benkumhene*–the chief in charge of the left wing of Biadan (15 December 2011)–who claimed to have engineered the formation of the Berekum Local Spiritual Assembly of the Baha'i Faith in 1991, he said that prior to this he had been a member of the Berekum Catholic Church, but had relinquished his Catholic faith in favour of the Baha'i Faith in 1988. He explained that in the Baha'i Faith, members are organised into what he calls 'Local Spiritual Assemblies' and 'Groups'. If a set of Baha'i has more than eight people, they are classified as a 'Local Spiritual Assembly', while they are classified as a 'Group' if they have eight members or less. He said that there is another Local Spiritual Assembly at Botokrom, and six other Groups in the traditional area. The Groups are found in the Biadan, Senase, Koraso, Jinijini, Amomaso and Mpatapo communities. According to *Nana*, there are currently 400 members of the Baha'i Faith in the traditional area, but they plan to increase their membership in the traditional area, and to do so, the Berekum Local Spiritual Assembly is inviting missionaries to augment their effort. And indeed, during the course of the interview, a young lady was introduced to me as a missionary who had just arrived in Berekum from the Volta Region of Ghana to begin working on this project.

The relationship between the Baha'i and the traditionalist communities

According to Nana Kuma Acheampong (Personal Communication, 15 December 2011), the Baha'i get on well with the traditionalists. Nana praised the traditionalists for their good attitude towards nature, especially for the institution of taboos to protect certain trees, animal and rivers, and described indigenous religion as an environmentally friendly religion. *Nana* was emphatic that in terms of their attitudes towards nature, the Baha'i Faith and African Traditional Religion are on the same path.

Moreover, Nana said that the Baha'i Faith believes in the value of nature and thus teaches its members to protect nature by refraining from indiscriminately killing animals or cutting down trees. According to him, the Baha'i Faith strongly discourages its members from setting bush fires as they cause serious damage to the environment. He also emphasised that the Baha'i Faith stands for peace and development in the world, and that this religion supports any move that promotes the management of the environment in a sustainable way. He stressed that the Baha'i Faith believes that the survival of humanity depends on human beings' attitudes towards the environment, and hence they encourage their members to be responsible in their use of it. He also mentioned that members of their faith are involved in tree planting exercises in the traditional area, and have planted trees around the Holy Family Hospital area in Berekum, as well as in the towns of Koraso and Jinijini. In addition to this, he said that they regularly organise clean-up exercises in Berekum town, particularly at the Holy Family Hospital premises. *Nana* also said that until her death, his wife had used the three local private radio stations in Berekum to encourage members of the communities to stop degrading the environment.

Summary

This chapter has focused on the worldview of the people under study. The worldview of the Berekum people was argued to comprise a set of assumptions that they hold about the universe which enable them to understand and interpret the nature of reality and to express this through the culture of the people. It was noted that worldview forms the basis of the people's conduct and way of life, and that the people's worldview helps them to live in harmony with their environment. It was also shown that the people's worldview is not static, and has an element of dynamism that enables it to accommodate the realities of the present. As the Berekum people's worldview dictates their conduct, it was found to have implications for their ecological practices as well. The discussion of this topic touched upon the religion of the people, since their worldview was shown to be largely underpinned by their indigenous religion. It was found that the spiritual elements of Berekum's indigenous religion are hierarchically arranged, with God heading the hierarchy. This discussion brought out the fact that the factors such as colonialism, Western education and the inroads of foreign religions in the study area (particularly Christianity) have affected the hold that indigenous religion has there, with many local shrines now dormant. However, it was also shown that Berekum's indigenous religion has displayed resilience in some areas, with beliefs in river and tree deities, the power of the ancestors and some lesser gods, and the fear of witches all remaining strong. Finally, both the decline and the resilience of indigenous beliefs were found to have ecological implications for the conservation of nature in the area.

PART III:

INDIGENOUS CONSERVATION STRATEGIES AND CHALLENGES

CHAPTER FIVE

BEREKUM RELIGIO-CULTURAL PRACTICES AND ENVIRONMENTAL CONSERVATION

Introduction

In this chapter, I present information derived from my interactions with my key informants in Berekum Traditional Society concerning their indigenous ecological practices. But before then, I briefly examine the basis of traditional African environmental code or ethics on which the people under study's environmental behaviour is based.

Foundation of African Environmental Ethics

African environmental ethics is derived from a worldview that is strongly theistic. Agboro (2008) argues that the majority of Africans conceive humans and their environment to be 'two inseparable entities that cannot be divorced from each other'. Nwosu (2010) corroborates this, contending that many Africans believe that without the environment, human life may not be sustained. Furthermore, he points to studies that show that African cosmology conceives of the existence of human beings as being tied up with their environment. According to Opoku-Ankomah *et al.* (2010), culturally acceptable environmental management among Africans emanates from social organisation that is permeated by spirituality and a reverence for ancestors. Dopamu (2003, p. 444) also argues that African peoples' ancestors addressed their environmental problems through the use of the 'available resources of nature to procure [the] non-therapeutic needs of man.' They applied the art of influencing courses of events by means of ritual behaviour and the invocation of divinities or spirit aids (Nwosu, 2010).

Ontologically, traditional African people believe that human beings did not just happen to be in this world, but rather that they were designed to live in the world and designed in relation to their environment(s). This implies that environmental consciousness is embedded in the traditional

African worldview (Awuah-Nyamekye, 2009a)–humanity, under such a worldview, is essentially connected with nature (Schulkin and Sarokin, 1996; Mawere, 2011). Abayie-Boaten (1998) stresses that because of this, many traditional African societies conceive of themselves as stewards of the environment, with social and moral obligations to manage and conserve it for posterity (see Attfield, 2010; Mawere, 2014). This social and moral obligation is a central part of African 'environmental ethics'.

Under this ethical code, the people of Berekum traditional society and in fact Africans in general often view land and its resources as a divine gift to the social group, or as communal property that the living hold in trust for their ancestors and future generations (Omari, 1990; Byers et al., 2001; Asiama, 2007). For scholars like Byers *et al.* (2001) and Asiama (2007), many African people see their relationship with the land as a matter of spiritual concern. The Akan, together with many other traditional societies in Ghana, view the land as an important deity (Parrinder, 1961: 38; Opoku, 1978: 56).

Fortes' (1945: 176) studies among the Tallensi of northern Ghana revealed that the people 'stand in awe of the earth and speak of it as a "living thing" which they believe can intervene mystically in human affairs as the ancestral spirits do'. Danquah, (1928: 212) outlines the Akyim of Ghana's conception of land as the property of their ancestors. For them, it is not the property of the living, but is something that they should maintain and preserve for future generations. Timberlake and Cunliffe's (2007) study of Zambezi society, and Nwosu's (2010) study of the Okonko society of Nigeria have produced similar findings. This suggests that living members of these societies only hold the land in trust for posterity, and do not conceive of themselves as having the right to sell it. Edusah (2008) quotes Danquah (1968) as having observed that this is grounded in the Akan conception of natural resources and the need for conservation, stating:

> .. in our culture, we do not only hold in trust for the present and future generations all the natural resources on which our welfare and [the] continuance of the community depend, but also are accountable to [them] for the proper management of the resources (cited in Edusah, 2008, p. 4).

This position contrasts with what the Christian Bible says in Genesis, 1:28, which Lynn White (1967) criticises. However, White's criticism is rejected on the grounds that it involves a misunderstanding of the text, as well as being overly simplistic as domination *implies* stewardship (Attfield, 1983, 2010; Warren, 1995). I am aware of the caution not to romanticise the role of religion and culture in nature conservation in indigenous or traditional societies; (Ellen, 1986, 1993; Brightman, 1987; Milton, 1996, 1998). However, the evidence on the ground is compelling in terms of the potential of religion and culture to influence traditional peoples in relation to the conservation of nature. For instance, the study undertaken by Byers *et al.* (2001: 205) in the sacred forests of Zimbabwe shows that 'the forest loss is dramatically less in forests that are now considered sacred, or were in the past connected to these sacred forests, than in forests that were not formerly connected to a current sacred forest'. In their analysis of the responses of their interviewees, they go on to state that the reasons that local residents give for the importance of respecting sacred forests are grounded in their perceived values together with the benefits they accrue from their uses of those forests. Furthermore, a study in parts of the Brong-Ahafo region of Ghana–where the current study is also located–revealed that despite the fact that local health and law enforcement officials have explained the health and legal implications of fishing with poison to the villagers there, it was the indigenous beliefs and taboos that effectively reduced the incidence of pesticide fishing in the area (Dyasi, 1985; Appiah-Opoku, 2005). This view supports my thesis that indigenous religion and culture have the potential to address some of the environmental crisis that Ghana is currently experiencing. This also calls into question the reductionist or simplistic hypothesis which hold that today`s environmental crises can only be effectively handled through secular means (science and technology). This re-emphasises the need for further research into indigenous religions and nature conservation in relation to the potential benefits that can be derived from this relationship. This, as I have pointed out, is one of the motivations for this book.

Berekum's indigenous ways of conserving the environment

In response to a question I posed to an ex-chief in Berekum regarding the philosophy behind the indigenous Berekum people's ecological practices, he responded by citing the proverb *'Adidaa yɛ kyɛn adiprɛko'*–'it is not good to eat all that you have in a day'. This statement can be seen as a summary of the motivation behind Berekum's indigenous ecological practices

In Berekum traditional life and thought, as in many traditional African societies (Danquah, 1928; Byers *et al.*, 2001; Schoffeleers, 1978), as I pointed out in the above, human beings are conceived of as being in a relationship with the other creatures of the universe. This implies that in the worldview of the people under study, gods, spirits, ancestors, spiritual and political leaders, lands, forests, rivers, and animals are all interrelated. That is, traditional Berekum people see their lives as interrelated and interdependent with other life forms, which mean that a holistic framework is needed to study Berekum environmental ideas and ethics. Informed by their worldview, the people of Berekum have put in place measures to conserve nature. The major ones that are examined in this chapter include: *Kwaebɛnnɔ (*sacred groves); the belief in *akyeneboa* (totem); a general attitude towards protecting A*saase Yaa* (land/earth), *nsuo* (water bodies), and landscapes; the existence of *nkyida or nnabɔne* ('hateful days' or 'taboo days') and the planting of *Ngyedua;*[1] and the belief in *Sasa* (A kind of *tumi* (spirit) believed to be residing a certain kind of flora and fauna).

It is important to note that the above forms of traditional ecological knowledge and practice are not as effective as they used to be for conserving the environment in Berekum as a result of a multiplicity of factors as discussed in Chapter Three. However, it is still relevant to discuss them, as they remain the main means for conserving nature in the study area.

The origin of the *Kwaebɛnnɔ* (sacred grove)

Berekum traditional worldview, as it is among the Akan in general, as I have argued (see Chapter Four) poses two dimensions to life. That is, the 'sacred' and the 'profane'. The word 'profane' here is not being used in the sense of moral judgment, but rather, in the sense as used by Eliade (1959). That is, the word is being used here to denote the visible, physical or the material dimension of life. It consists of our physical environment and the secular activities that go on in it. The 'sacred' on the other hand, is used to designate the invisible, immaterial or the spiritual dimension of life. The sacred universe in the view of the traditional Akan people is made up of personal spirit beings, both human (ancestors) and non-human (gods). There are also impersonal or non-moral forces which manifest themselves in the working of magic, witchcraft and sorcery. Although the people seem to make a distinction between the spiritual and the physical, this distinction as I have pointed out, is artificial because they at the same time firmly believe that these worlds are not separate entities but one (see Olupona, 2001), with the sacred or the spiritual showing itself through the

[1] *Ngyedua* are large-leafed tress of the ficus family usually planted to serve as *genius loci* for the community, windbreak and for the beautification of the community.

physical. Again, in spite of the fact that the sacred and the profane or the physical are seen as one entity, the word 'sacred' is usually reserved for things that they view as having awe-inspiring or divine attributes.

Based on this conception of the universe, when the people of Berekum traditional society designate a place/space as 'sacred', they mean that place is capable of revealing that which is 'holy' and should be considered as such. That is, that space has or is imbued with religious significance. This means that among the people under investigation, a distinction is at times made between an ordinary space and that of a 'holy' or sacred space. This in a way may be said to be similar to how the Australian aborigines assign sacred or profane status to things as Durkheim's (2008) study among them revealed.

The sacred grove is a well-known phenomenon in Africa and elsewhere, particularly in various parts of Asia (Laird, 1999; Chouin, 2002, Fomin, 2008; Ntiamoa-Baidu, 2008, 1995). However, their history and evolution is unclear. Whilst some believe them to have originated relatively recently in historical terms, others think that some of these groves 'may be as old as humankind' (Kingdom, 1989, quoted in Attuquayefio and Fobil, 2005). Thus, varied historical accounts are connected with the various sacred groves that are found around the world. Sacred grove in the study area is known as *kwaebɛnnɔ*, a word that is constructed from two Akan words of the Bono dialect–*kwae*, meaning 'forest', and *bɛnnɔ*, meaning 'that which is not weeded'. In an interview with Nana Akumfi Ameyaw (7 November 2011), he explained that the word *kwaebɛnnɔ* derived from the fact that those who have cultivated or attempted to cultivate such forests have either died or suffered misfortunes of some sort. Such beliefs led people to conclude that, 'we do not weed this forest', hence *kwaebɛnnɔ.*

These groves are known by various different names among the Akan of Ghana, including *nananom mpow* (ancestral groves) (Adakwa-Dadzie 1997), *asɔnoyɛso/abosompow* (shrines), *mpayinpow* (ancestral forests) and *nsamanpow* (burial grounds) (Ntiamoa-Baidu, 2008). *Abusuapanyin* Yaw Nana Nkrumah (Personal Communication, 13 November 2011, Akroforo) explained that these groves were created by the forebears of their community long ago as a way of preserving nature. He explained that although most of these *kwaebɛnnɔ* are relatively small, they nonetheless provided a modest means through which *nananom* attempted to protect and conserve nature for posterity. Studies on sacred groves across the world reveal that they are normally small, although some are quite large forested areas found near settlements (Kingdom (1989). Access to sacred groves is usually governed by strict taboos and customs (Lebbie and

Guries, 2008; Mawere, 2014), and research has estimated that there are between 2,000 and 3,200 sacred groves scattered throughout Ghana (Tuffour, 1991; Gordon, 1992), constituting about 1% of the forest that remains outside existing reserves (Fetzner, 2007). In addition, most of the origins of sacred groves are connected with the religio-cultural beliefs and practices of the community concerned (Ntiamoah-Baidu, 2008). This is the case in the study area, with the histories of most of the groves there being linked to supernatural forces. The history of two of these groves–*Mfensi* and *Nyina Kofi*–will now be briefly presented, as representative illustrations.

The Mfensi Kwaɛbɛnnɔ

Mfensi sacred grove is one of the most important sacred groves in the study area. It is located in Kato, a village lying about 1.5 kilometres from Berekum, the capital of the Berekum Traditional Area. It is found at the south-eastern part of Berekum, and covers approximately 40 hectares of land. It appears to have not been disturbed for a very long time, although there were allegations approximately three decades ago that some of the elders in the village had connived with a Syrian timber contractor based at Kato to fell some of the trees in the grove, and these rumours caused an up-roar in the village (Nana Siraa Ababio III and her elders, Personal Communication, 7 December 2011, at her palace in Kato).

Figure 13 . Entrance to the Mfensi sacred grove

Nana Afia Siraa Ababio III, the Ɔhemaa (loosely translated as Queen-mother) of Kato told me that when their ancestors first settled in the area as farmers, they decided to cultivate the place where the grove is now located in order to make farms. However, every time they decided to cultivate the land, they would hear a strange voice saying '*mfa nsi hɔ*', an Akan sentence meaning 'do not put it there'. Moreover, she said that those who entered that section of the forest to set up traps (or for any other reason) would be warned '*Mfa nsi hɔ'*. This led the people come to fear the place, and to view it as the abode of a very powerful *sunsum* (spirit). This *sunsum* was named *Mfensi*, a name deriving from the warning that the spirit gave to those who attempted to encroach upon it. As a result, this section of the forest, thus, became a 'no-go' area in the community, and later became a sacred forest for the people of Kato. The grove has been protected through unwritten regulations that were put in place by the elders of the community. A shrine was built within the grove, where the attendant appointed to oversee the management and protection of the grove must go from time-to-time to pour libation to *Mfensi*, the guardian spirit of the grove. The grove consists of tall trees that form a canopy with lianas, and dense undergrowth through which the *Bɔgyesɛ* stream flows.

Figure 14 The Mfensi grove's attendant offering a libation prayer in the presence of the author and Dr. Kevin Ward.

Nyina Kofi Kwaebɛnnɔ

This grove is located at Koraso, a village that lies on the north-western part of Berekum, about 5 kilometres from Berekum town. The grove was created because of an experience that a hunter called Nana Kofi *Pambоɔ* had in that part of the forest during one of his hunting expeditions. According to the *Nyina Kofi Bosomfoɔ* (the priest in charge of the grove)

Figure 15.The researcher in an interview with the Nyina Kofi Kɔmfoɔɔ within the grove.

Nana Kwame Agyemang, when the hunter Nana Kofi *Pamboɔ*, their great ancestor, got to the spot where a large *onyina* (*Ceiba pentandra*) tree stood, he saw a lot of gold scattered under the tree. He took some of it, but then could not find his way back home. When he put the gold back from where he had taken it, he was able to find his way home again. According to the priest, he tried several times to take some of the gold home, but could not succeed. On returning home, he informed his elders about his experience in the forest. He was advised to consult an oracle for an explanation. When he did so, the answer he received was that the gold belonged to a powerful spirit that lived in the forest, and if the people would be ready to serve it, then the entire village would gain from its protection. He informed the elders of the community about what the oracle had revealed. They accepted it and the necessary rituals were performed. After the rituals, that portion of the forest was designated as a sacred grove, and a priest was appointed to be in charge of it. The grove was named after the hunter Kofi Pamboɔ and the *onyina* tree under which he had the experience. Hence, the name *Nyina Kofi Pamboɔ* sacred grove–although the *Pamboɔ* part of the name is not often mentioned. *Kofi* is also a name given to an Akan boy born on a Friday, which suggests that it is believed that the protective spirit of the forest is a male.

According to the priest, the promised help from the spirit came during the Berekum-Koraso war in 1930/31 when, despite the numerical strength of the Berekum people, the people of Koraso could not be defeated. Their success during the war was attributed to the spiritual support they received from *Nyina Kofi*. The priest mentioned that after the war, and to their surprise, they discovered that several bullets had been gathered under the *onyina* tree, and sap/juice was oozing out of the tree like blood. Upon oracular consultations, it was revealed that the bullets were those fired by their attackers during battle, which explained why they were not defeated during the war. The grove is thus very important to the Berekum people.

Another important *odum* tree in the grove is known by the local people as *Dum Abenaa* (literally 'a female *odum* tree born on a Tuesday'), and this tree has a small shrine under it. *Dum Abenaa* is believed to be the wife of *Nyina Kofi.*, which again shows the anthropomorphic nature of Berekum indigenous religion. There is an annual pilgrimage to the grove for rituals, which precedes the annual festival of the people of Koraso. Near the *Nyina Kofi* grove can be found the water head of the River Koraa, the main water supply for the entire traditional area. It is important to stress that a lot of sacred groves are scattered in the study area, but only those close to the towns and villages are known to all the members of the towns or villages. Those further away from the towns and villages are known only to those who farm close to them. This implies that more of the sacred groves would be discovered if a lot more of the people were interviewed.

Figure 16 The tall tree shown in the photo is the Dum Abenaa, the supposed wife of Nyina Kofi

From the interviews, I was able to compile a list of some of the sacred groves in the study area. I did not only ask for a list of them, but their location, the spirit being(s) associated with them, and their current status. I visited most of these groves and took photographs of the sites. Some of them were videoed. Information about the ones I found out about is presented in the following table.

Table 8. Some sacred groves in the Berekum Traditional Area

Community	Name of grove	Location/sacred site	State of grove	Associated Spirit (s)			
				Gods	Ancestors	Dwarfs	Legends
Biadan	*Adwamɛ*	Kuromo	Wiped out	X	X		
	Bokyɛ mso	Near River Koraa	Degraded	X		X	
	Dummonta	Asuotiano road	Destroyed	X	X	X	
	Berekum kɛse	Hiaa AreaAlong the Koraa	Degraded		X	X	
		River	Preserved but			X	
	Koraa Banim		isɔlated				
		Asutia Road		X	X	X	
	Wamkonagya		Encroached				
		Far away from the village		X		X	
	Kwangyemire		Degraded				
				X			
Akroforɔ	*Boɔ kwaɛ*	Near the village	Reduced	X	X		
	Wuo kwaɛ	Near the village	Reduced	X	X	X	
	Nyinanatiamu	Near the village	Reduced	X			
KutreNo. 1	*Boɔ Kwae*	Near the village	Destroyed	X		X	
Mpatapo	*Owuo kwae*	Near the village	Reduced	X	X		
Koraso	*Nyina Kofi*	Near the village	Preserved	X	X		
Kato	*Mfensi*	Along the Berekum-	Protected	X	X	X	
	Mpɛm,	Sunyani highway	Reduced		X		
	Bɔtiso	*Kwadwofoɔ bɛhu* road	Reduced		X		X
	Boɔkofi	Near*Kwadwofoɔ bɛ*	Reduced	X	X	X	
	Nsereso	*huAdesu* area	Reduced	X			
		Kwadwofoɔ bɛhu road					
		Kwadwofoɔ bɛhu road					

Pepaase	*Nananompɔ* (Berekum Royal Mausoleum)	Along Pepaase-Kotaa road	Protected	X	X		
Mpatasie	*Kontonkori* *Mangyina kwae*	Near the village Near the village	Destroyed Substantially reduced	X 	X X	 X	
Jamdede	*Boɔ bɔn* *Asuotiaso* *Boamaso* *KoraasoAsɔno yɛ so*	Near the village Near the village Near the village Near the village	Protected but reduced Encroached Destroyed Protected	 X X	X X 	X X 	 X
Domfete	*Bɛntim kwae*	Along Domfete-Jinijini road	Substantially reduced		X	X	
Senase	*Adeɛmmra kwae* *Kwaabaa kwaebennɔ* *Koraa asɔnoyɛso or Banim* *Atwiso kwaɛbɛnnɔ*	Near Akatim Near the village Along the Koraa River Far away from the village	Reduced Destroyed Protected Reduced	X X X X	X X X	 X X	
Nsapor	*Asuonsapo kwaɛbɛnnɔ*	Along Nanasuano road	Encroached	X	X		

Source: Author's Fieldwork Notes, 2012

An important feature of the sacred groves in the study area, is that restrictions are placed on their entry. These are regulated by local taboos, and most of the groves have accredited attendants, who are usually traditional priests. These are the people who can enter the groves regularly for religious purposes, and do so on the sacred days dedicated to the relevant groves. The grove attendants I interviewed said that the groves also act as *asɔnoyɔso* (shrines), where people can go for supplication, to redeem pledges, and to remove or overturn the effects of *dubɔ* (curses/grievance imprecations). Even when people are allowed to enter the groves, they are only allowed to enter certain parts, which exclude what are referred to as the 'holy of holies'–usually small areas within the groves that have been prepared to serve as the actual shrines. It is here that ordinary people (worshippers) are not allowed to enter, and those who visit the shrines within the grove are not allowed to touch anything there, although on rare occasions medicine men/women are allowed to take some herbs that are difficult to find outside the groves. Usually special permission must be given from the attendant before this can be done, and sometimes the attendant must check with the chief of the town or village before he or she can grant permission for herbs, bark or roots of plants to be taken from the grove. This is to prevent the over-exploitation of herbs. The restrictions that have been placed on entering groves have meant that they have provided sanctuaries for wildlife.

One important issue that I was interested in pursuing during the data collection from my key informants was whether the creation of sacred groves was informed, first and foremost, by conservationist or religious motives. I was keen to follow this up due to the debate that has emerged in academic circles regarding 'whether conservation is a deliberate purpose of such groves or only a fortuitous coincidence..' (Lebbie and Guries, 2008: 43. See also Githitho, 2003; Tomalin, 2002; Sarfo-Mensah, 2001). I, therefore, posed the following question to all my informants: *Some people are of the opinion that the conservational value of sacred groves in indigenous societies is mere coincidence, for their creation was influenced primarily by religion rather than by conservationist motivation. What do you think about this?* The recurring response from my key informants was that the creation of the sacred groves and the methods they have in place for dealing with environmental issues have a *primarily* conservationist motivation rather than a religious one. In other words, their attitude towards nature is informed by their quest to conserve and protect natural resources. They said this resulted from the experiences that the forebears of their communities had acquired.

This seems to contradict the empirical evidence on the ground. This is because the histories of all the sacred groves I visited had spiritual or religious beginnings. In addition, various stories were told at many of the communities I visited about the misfortunes that befell those recalcitrant individuals or groups who violated the taboos associated with places that have been designated as sacred groves or sacred places. There was only one exception in which no religious connection could be found for a sacred grove, which was the *kwangyemire* grove that lies well away from Biadan town. I was informed that this grove was purposefully set aside for hunting, and there were rules and regulations in relation to how hunting could be performed there. Even here, one cannot wholly say that the *kwangyemire* grove was 'set aside' for conservational purposes, then. I was specifically interested to know or see a grove that had a clear conservationist motive for its beginning, but I could not find one.

I tried to understand why what my informants were telling me was different from the evidence on the ground. Nana Ansu Gyeabour (Personal Communication, 8 November 2011) explained that the conservationist idea first came from the chiefs and the elders of the various communities, who asked for the support of the *nananom, abosom* and other *asunsum* (spirit beings) of the community. For this reason, the community immediately designated a place as a *kwaɛbɛnnɔ*, and prayer, libation, sacrifice and other rituals were performed to sacralise it. According to *Ɔkɔmfoɔ* Agyemang, this sacralisation resulted in the population of the place(s) by the various spirit beings that punish non-conformists in the communities (Personal Communication, 14 October 2011, at his shrine in Jinijini). This explanation was also corroborated by some chiefs.

My informants were unanimous that religion was used to serve as an *enforcer* of conservation. That is, religion was used to ensure the effective enforcement of the rules and regulations made for promoting environmental conservation. In the life and thought of the people of Berekum traditional society, as in many traditional African societies, religion is such a powerful institution that issues that have religious connotations are treated with respect and awe. Therefore, attaching religion to conservational rules is one of the surest ways to make people respect and abide by rules and regulations, even in solitude. In other words, religion is used as a vehicle for achieving conservation goals. This also explains why in this society the line between their religious ans secular lives is very thin.

My informants also grounded the existence of Berekum conservationist measures on the idea that they must show gratitude and respect for the Earth's resources for their own good (Korang-Amoakoh, personal communication, 15 October 2011, Berekum). This suggests that the people of Berekum are aware of the popular saying that 'when the last tree dies, then the last man also dies'. In other words, they seem to be saying, to borrow the Kenyan activist and 2004 Nobel Peace Prize winner, Wangari Maathai's words, 'In the process of helping the earth to heal, we help ourselves' (2011).

It is important to point out that some of my informants reacted very strongly to the suggestion that the conservational values of sacred groves are mere coincidence. Nana Akumfi Ameyaw (personal communication, 7 November 2011), for instance, sums up the views of those who disagreed with the above suggestion:

> Conservationist knowledge has been with us [Africans] for a very long time, before *Kwasi Broni* (Whiteman) arrived in Africa. This is evidenced in our everyday speeches, proverbs, folktales, myths, etc. We vehemently reject this suggestion. If we did not have conservationist knowledge, we would not be socialising our children on the need to make use of natural resources in a sustainable way. This is a neo-colonialist view.

In fact, I saw evidence during my interviews that confirmed the above claim. For instance, I observed that from infancy in traditional Berekum society, the individual is socialised to see land as a sacred entity as he/she grows. Moreover, all the techniques for exploiting the land and other natural resources in a sustainable way are part of the socialisation programme among the people. This makes it difficult to rule out conservationist motives for the people's actions and to argue that the ecological values resulting from such actions are simply a by-product of religious motives, as the 'skeptical religious environmentalists' do.

From the above, it is however, possible for the casual observer to conclude that it is religion rather than conservationist motives that underlie indigenous people's actions. However, I argue that this is a simplistic view on indigenous African people's relationship with nature--as evidenced in this case study--this is because the hold of religion on the people is so strong that it becomes difficult for one (especially the casual observer) to make a distinction between their secular and religious activities. The clearest way to understand their conservationist drive/attitude is to analyse their everyday speech, particularly their proverbs. As I have pointed out earlier on among the traditional Africans, the major law enforcement agent is their religion. Religion is viewed with awe that nobody is prepared to violate any law that has religious underpinnings. It is this that makes the casual observer to conclude that the conservational value of indigenous peoples' actions is just a by-product of their religious behaviour.

The ecological importance of sacred groves

I have already noted that many scholars see sacred groves as being important for conservation, but whether their creation was motivated by conservationist motives has become the subject of scholarly debate. Using the Shona of Zimbabwe as a case study, Taringa (2006) argues that the influence of indigenous peoples' worldviews on human-nature relationships 'are primarily relationships with spirits and not necessarily ecological relationships with nature' (Taringa, 2006, p. 196). Closely connected to this is the debate concerning whether, due to their small size, sacred groves could be of any real ecological significance or value. Tomalin (2013) quotes Freeman (1994) and Kalam (1996) as drawing attention to the fact that sacred groves 'are not necessarily evidence of "environmentalism", and second, that there is not an ideal type of sacred grove that is constituted by "pristine forest"' (quoted in Tomalin, 2013, p. 199).

But scholars such as Riiters, Wickham, O'Neil, Jones and Smith (2000) and Terborgh (1992) have acknowledged that research undertaken by Guindon (1996), Lugo (1995), Nkongmeneck *et al.* (2002), and Turner (1996) have shown that 'the level of diversity remaining in secondary and fragmented forests (sacred groves) still makes their conservation important as they may become the sole sources of native species for future restoration efforts' (Lebbie and Guries, 2008, p. 43). Because of this, I requested information from Mr. Asare (the forester attached to the Asuo Bomosadu Timbers and Sawmill (ABTS) Company in Berekum) about the types of tree species that are now extinct in the study area during my data collection phase. He listed the following trees: *Odum* (*Millicia excelsa*), *Kokrodua* (*Pericopsis elata*), *Papao* (*Afzelia Africana*), *Kusia* (*Nauclea diderrichii*), and *Sapele* (*Entandrophragma cylindricum*). He attributed their disappearance to over-exploitation by timber firms and illegal chain-saw operators (loggers). Interestingly, many of these trees were identified by the priests or attendants in the groves I visited. I also observed many species in the groves that were abundant when I was a child, but are now significantly reduced in number in the area; and not only trees, but also several species of insects and butterflies (among other animals), which are important agents for pollination and seed dispersal. This makes sacred groves indeed, a potential sanctuary for flora and fauna.

Although table 8 (in this chapter) shows that the number of sacred groves in the study area has been significantly reduced, and that some of the existing groves have been encroached on for various reasons–just as Barre, Grant and Draper's (2008) study among the Tallensi-Nandam people of northern Ghana showed – the possibility of the cumulative worth of the remaining groves for the local environment should not be dismissed (see Amanor, 1994; Gyasi, 1997; Arhin, 2008). For instance, Palmer (1992) points out that even 'one hectare of forest may have between 100 and 200 hundred tress species [..] the vegetation feeds and provides shelter for huge number of invertebrates and large animals'(1992: 83). I also found that numerous groves in the study area have rivers and streams flowing through them, with many of these being situated where the heads of rivers are located. In addition, I saw evidence from the ground that most of the sacred groves serve as places where traditional medical practitioners fetch medicinal materials that are rare in the communities. For instance, *atwerɛnantam* (literally, "under the frog's toe"), a useful plant for medication in the area, was found in the *Mfensi* grove.

Another interesting finding noted by my informants is that people in the communities that have their houses close to groves do not have their roofs ripped off during heavy storms, as others communities whose groves are located far away from their settlements do, apparently because the tall trees in the groves act as windbreaks. In reference to the ecological importance of sacred groves in Ghana, Agyarko (2001) argues 'these sacred groves, as they are called, have various underlying beliefs and prohibitions, but the common denominator is that cutting of trees for timber is prohibited. These groves add considerable value to the protected area of forests of high genetic value' (2001, p. 7).

A numbers of studies have argued for the ecological importance of sacred groves. For instance, the Buabeng-Fiema sacred grove in the Brong-Ahafo Region is a sanctuary for more than 500 Campbell's Mona and 200 Geoffrey's Pied Columbus Monkeys, as well as numerous other species of animals and plants (Akowuah *et al.*, 1975; Fargey, 1991; Attuquayefio and Gyampoh, 2010). Ntiamoa-Baidu (2008) study of marine turtles and birds in coastal Ghana also indicates that within those communities in which turtles and birds are protected by local taboos and beliefs, the population density and the sizes of the animals are greater than in those areas in which they are used as food (ibid.). Similar findings are noted in other studies in Ghana (Dorm-Adzobu *et al.*, 1991; Decher, 1997) and other parts of Africa. For instance, Byers *et al.*'s (2001) study of sacred forests in Zimbabwe shows that 'the amount of forest cleared is dramatically less in forests that are now considered sacred, or were in the past connected to these sacred forests, than in forests that were not formerly connected to a current sacred forest' (2001:. 5). Nwosu's (2010) study among the Okonkwo society of eastern Nigeria elicited similar findings.

Gordon (1992) suggests that sacred groves could serve as the 'basis for environmental and public awareness campaigns' in Ghana, and it was on this basis that Decher (1996) contended that the continued conservation of sacred groves (and perhaps all natural habitats and wildlife in West Africa) will ultimately depend on an understanding and appreciation of their ecological functions. The ecological importance of sacred groves has also been acknowledged by global bodies like the IUCN–the world's largest conservation network–and UNESCO. UNESCO stresses that:

> Sacred groves have served as important reservoirs of biodiversity, preserving unique species of plants, insects, and animals. Sacred and taboo associations attached to particular species of trees, forest groves, mountains, rivers, caves, and temple sites should therefore continue to play an important role in the protection of particular ecosystems by local people.

It is important also to note that UNESCO has not just recognised the potential of sacred grove in biodiversity conservation, but has concretely gone into partnership with some local communities worldwide to ensure the preservation and conservation of such sacred sites, and to document both these community or indigenous-based conservation models and the biodiversity of these sites. Two of such UNESCO projects are the '*Jaagbo* sacred grove' in the northern Ghana under the auspices of the Cooperative Integrated Project on Savanna Ecosystems in Ghana (CIPSEG). The sponsorship was from 1993 to 1996. A preliminary study from the project revealed that the grove and its buffer zone had approximately 220 plant species, in comparison to 60 in outlying areas (Corbin, 2008). The *Anweam* sacred grove in the Esukawkaw Forest Reserve is another UNESCO project in Ghana. This suggests that the ecological importance of sacred groves cannot be overemphasised.

Akyeneboa (Totem)

The Berekum word for the phenomenon of totemism is *akyeneboa*. This Akan word literally means 'an animal that one leans upon or relies on for spiritual support or inspiration'. Although *akyeneboa* is defined as being connected with animals, plants and other things are also included in their conception of totemic objects. Therefore, in traditional Berekum society, *akyeneboa* is used to refer to the animals, plants, symbols, emblems or images representing a clan or a group of people.

Totemism has been the focus of much discussion within anthropology and religious studies (McLennan, 1869-70; Frazer, 1910; Freud, 1912-13;

Evans-Pritchard 1989 [1965]; Rose *et al*., 2003; Durkheim, 2008 [issued]; Mawere, 2014: 20-21). Some of the key issues of discussion have revolved around the distinction between civilisation and savage, and between culture and nature (Rose *et al.,* 2003:. 6). Secularisation and the Darwinian theory of evolution by natural selection have also influenced the discussion. The phenomenon of totemism is conceptualised as 'a relationship of mutual life-giving between human beings and natural species (or rarely, other natural phenomena)' (Rose *et al*., 2003: 5).

The term 'totem' derives from the North American *Ojibwa* word, *ototomen* meaning 'maternal relatives'. Many scholars have contested the validity of using the term 'totem' in other cultural contexts, since the term is derived from a specific North American *Ojibwa* word (Rose *et al*., 2003, p. 5). Rose *et al.* (2003) stress that the terms 'token' and, in many parts of Australia, 'dreaming' are used interchangeably with 'totem'. (Rose et al., 2003: 2). Some scholars even consider the word 'totem' to be offensive. Nayutah and Finlay (1988) are cited by Rose *et al*. (2003: 2) as maintaining that the term 'totem' is still used by some Aboriginal people in New South Wales today. It is interesting to note that the Aboriginal people of New South Wales' understanding of the phenomenon of totemism can be said to be similar to that of the traditional Berekum people, and that of the Akan of Ghana more generally.

In the words of Parrinder (1961: 172), totemism may be viewed as 'the use of images and emblems by groups and clans to represent some animal or plant after which the clan is named'.Smith-Asante's (2002) study points out that the *Bakwena* and the *Batloung* ethnic groups of Botswana are named after *Kwena*, the crocodile, and *Tlou*, the elephant, respectively.

Among the people of Berekum Traditional Society, the origins of *akyeneboa* are expressed in myths, perhaps to help people to remember the facts, which concretely manifest themselves in rituals, ceremonies and festivals. My informants explained that the relationship between a group and its totemic animal or plant can result from:

a) Their ancestors tracing their origin to that animal or plant;
b) The *akyeneboa* helping the forbears of the ethnic group, or the forebears benefitting from it in a special way, or;
c) The totemic object disclosing itself (through revelation) to the head (or other members) of the ethnic group concerned (see Awuah-Nyamekye, 2012c).

It is this close intimacy that Rose et al. (2003: 1) view as a kind of kinship and caring 'between people and particular parts of the natural world' or, simply put, a 'kinship with nature' (ibid.: 5). These *akyeneboa* are important to consider, because their status confers an ethical position upon a number of animal and plant species. It is, therefore, no surprise that many taboos have been designed by traditional Berekum people to ensure that acceptable or responsible relationships with their totemic objects are maintained, thereby conserving nature. Thus, this human-nature relationship that exists among the people of Berekum portends the ecological dimensions of the phenomenon of totemism, mirroring Strehlow's (1970) findings in his study of totemism among the people of central Australia.

There is also a special relationship between a society or a community and its *akyeneboa*, which enjoins a community or the people to protect their *akyeneboa* as it is part and parcel of the group (This is a commonplace belief in the traditional Berekum society;see also Awuah-Nyamekye, 2012c). Ali Mazru (1986: 50) puts this idea succinctly: 'their readiness to see a little of themselves and a little of their God in their surroundings made them shrink from abusing totems to satisfy an appetite'. One of my interviewees explained that historically, when one accidentally killed a totemic animal, one needed to organise a funeral for it, just as if a human being had died. The *Onyini* (python) was mentioned in connection with this tradition in the study area. This protection ensures that the kinship relationship between people and their totemic object is maintained. For instance, many of the farmers I interacted with stressed that when they are preparing a new farm and their totemic plant happens to be where the farm is being made, care is taken so as not to burn the plant(s) during the burning stage of the farm. This demonstrates the kind of relationship between the people and their totemic objects. Nana Ameyaa (Personal Communication, 15 October 2011) observed that the destruction or extinction of a family or clan's *akyeneboa* implies the extinction of the family or clan concerned. Thus, the protection of *akyeneboa* is an important obligation for the people of the study area.

Trees that are believed to be spiritually powerful are usually found to be *akyeneboa* of Berekum clans and families, and thus they (and other clans within the Akan) view many plants, shrubs and climbers as totems. The origin of the *akyeneboa* and their presence among the Berekum and the Akan is generally thought to be restricted to the religio-cultural sphere. However, it is obvious that their attitude towards *akyeneboa* also underpins the sustainable use of natural resources in the traditional milieu. In this context, we can see how a worldview has been translated into practical ethical conduct through a set of simple actions. Thus, the application of totemism has become one of the major traditional strategies for the conservation of flora and fauna among the traditional Berekum people. In other words, the use of *akyeneboa* among members of the traditional Berekum society provides a framework for respecting and conserving nature.

In the traditional Berekum society, every individual must belong to one of the eight clans of the Akan. A major characteristic of these clans is that each has its own *akyeneboa*, which means that each person in Berekum must have a major *akyeneboa* in addition to the *akyeneboa* of his or her individual families. Table 9 below lists the Akan clans and their respective *akyeneboa*.

Table 9. Akan Clans and their Totemic Animals

Clan	Totem/Local Name	English Name	Scientific Name	Symbolic Quality
Aduana	*ɔkraman*	Dog	*Canis domestica*	Skill
Agona	Ako (awidie)	Parrot	*Psittacosis erithacus*	Eloquence
Asakyiri	*ɔpɛ tɛ*	Vulture	*Neophron monochus*	Cleanliness
Asenie	Apan	Bat	*Myotis lucifugus*	Diplomacy
Asona	*Kwaakwaadabi*	Pied crow	*Corvus albu*	Wisdom
Bretuo	*ɔ sebɔ*	Leopard	*Panthera pardus*	Aggressiveness
Ɛ kuɔ na	*Ɛ koɔ*	Buffalo	*Syncerus caffer*	Uprightness
ɔy ɔko	*Akrɔ ma*	Falcon/Hawk	*Falco columbarius*	Patience

Source: Adapted from Conservation International, 2003: 2.

Aside from totemism, the people's beliefs in certain other phenomena are also used to conserve nature. This includes their beliefs about the earth, water bodies and unusual landscapes, which I will now discuss.

Asaase Yaa (Land/Earth)

The concept of the earth or the land among traditional Berekum and traditional African societies in general derives from their worldviews. During an interview, one Berekum Chief observed that '*Asaase no ye nananom ne nkyirimma dea yɛn deɛ yɛ rehwɛ so kɛkɛ*', which translates as 'the land belongs to the ancestors and the generation yet unborn, we are only care takers'. This is similar to something that Danquah (1928) said about the Akyim, that the land belongs to the ancestors and future generations, but is not the property of the living. The sacredness of land among Africans is also something that Dikirr (2005) found in his research among some African people.

The value that traditional Berekum people attached to the land led them to view it as a spiritual entity and goddess, with the name *Asaase Yaa* (Earth Yaa). *Yaa* is an Akan name given to a female girl born on a Thursday. This explains why the Earth is known as *Asaase Yaa* in the study area, as well as in many other Akan societies. Explaining the femininity attached to the earth, Nana Ameyaa said that 'the land nurtures human beings like the way women do to their children' (Personal Communication, 15 October 2011). This emphasises the anthropomorphism in Berekum traditional religion.

Assigning femininity to the earth is not peculiar to the Berekum people, but is a common practice in traditional Ghanaian societies, as Asiama's (2007) study finds: 'land ownership in Ghana is characterized by religious beliefs and practices. […] To the Akan, land is a supernatural feminine spirit with Thursday as its natal day, which can be helpful if propitiated and harmful if neglected' (2007: 5).

Among the Berekums, *Asaase Yaa* and her resources are viewed as a divine gift to the social group or communal property which the living must hold in trust for their ancestors and the future generations, therefore, the individual is socialised to accept this view. Other studies among indigenous communities have generated similar findings (Sarfo-Mensah and Oduro, 2010; Asiama, 2007; Byers *et al.*, 2001; Appiah-Opoku and Hyma, 1999; Omari, 1990; Busia, 1968). It is thus unsurprising to find that land is not a sellable commodity among the people of Berekum Traditional Society. The only circumstances under which a piece of land can be sold is when this is necessary for solving an urgent problem, such as the defrayal of stool debt resulting from land litigation with other communities (see Busia, 1968). It is important to state, however, that although an individual may acquire land and have rights over its usage during the course of her life, she still does not own the land as such, but rather simply enjoys its usufruct.

In view of the sacredness attached to the land by the Akan and the traditional Berekum people, certain actions are considered as *hateful* to *Asaase Yaa*. These include tilling the land on a Thursday (*Asaase Yaa*'s natal day), murder (spilling innocent blood on the land), having sexual intercourse in the bushes, giving birth in the bushes, burying a pregnant woman without removing the foetus, and digging a grave without offering libation to the land first (see Awuah-Nyamekye, 2009a: 265). All these taboos are common knowledge in the study area. Busia (1968) and Opoku (1978) have made similar observations in their studies among the Akan. All these acts are deemed to be a defilement of the land, which may result in drought or infertility of the land unless purificatory or propitiatory rituals are performed–a point that all the Chiefs and traditional priests that I interviewed stressed (see also Frazer, 1926; Sarfo-Mensah and Oduro, 2007).

The traditional people's conception of land in the study area and throughout Ghana has resulted in people holding matters concerning land very dear to their hearts, and they are prepared to go all-out to defend their ancestral lands as this is viewed as a sacred duty. The spiritual connectedness to land has been seen as invoking extreme conservatism towards its protection and management (Mufeme, 1999). However, others argue that this has made the ownership and distribution of land in Ghana problematic (Goldstein and Udry, 2008; Amanor, 2001), affecting all policies in Ghana concerning it (Awuah-Nyamekye and Sarfo-Mensah, 2011). This is sometimes seen as anti-development as it makes the acquisition of land for development difficult, which has led some to call for radical land reforms in which the central government would take full control over all the land. My key informants oppose this on the grounds that the sacredness and the sanctity attached to land would be lost if the government (which is influenced by secular views) took control of all the land in the country.

Nsuo (Water bodies)

In view of the crucial role that water plays in human life, the people of Berekum Traditional Area have put in place certain measures to protect water bodies. For instance, it is a taboo to defecate near water bodies, particularly rivers, or to farm close to river channels Awuah-Nyamekye, 2009b). These restrictions are used to protect rivers. Sarfo-Mensah, Fredua Anto and Oduro's (2010) study among other Akan people has confirmed this finding.

For both Berekum and Akan people in general, rivers, lakes, lagoons, springs, and ponds are viewed as sacred entities, and are considered to be *nsuo abosom* (river/water deities). *Asuo Koraa*, (The River Koraa) is the principal river in the study area, and runs through several towns and villages there. The local people consider *Asuo Koraa* to be a female, and the fishes in it are seen as her children. Fishing is, therefore, strictly banned in *Asuo Koraa*. Those who flout this injunction have suffered for it according to local stories. An account from a joint interview during the data collection phase of this research is presented below:

> I quite remember some time ago, one Anthony–a Syrian timber contractor at Kato–doubted this by saying that it was just an "African belief", and ate some of the crabs. After that he felt that something like one of the hands of the crab has choked his throat. He went to hospital for treatment but the problem persisted. He was even sent out of the country for medical operation but could not get well. There were series of operations but the medical doctors could not diagnose the source of the problem. Finally he came back to Ghana to confess eating a crab from the River Koraa. He had to pacify the river goddess before he got healed. My own father was involved when they were performing the rituals which restored Anthony's health (Nana Ansu Gyeabour, personal communication, 8 November 2011).

This belief in river deities is still very strong among the people. Several *asɔnoyɛso* (shrines) can be seen dotted along the course of *Asuo Koraa*, and the land that it flows through is considered as a sacred grove. Friday is designated as the sacred day for *Asuo Koraa* and, on this day, all her designated attendants go to the shrines along the course of the river to offer her libation prayers and food. As part of the activities, people from all over the area come to visit the shrine to present their petitions or to redeem vows they previously made to the river goddess. Fishing is not common in the Berekum area as there are few other big rivers, and those who fish in the small rivers have to use appropriate methods. For example, it is a taboo to use any chemicals for fishing. All the rivers in the study area are treated with a degree of reverence, but this is gradually diminishing (as a result of the influences discussed in Chapter Three). However, one may argue, based on the foregoing discussion, that the traditional Berekum people's attitude towards water bodies–particularly the way in which they ensure the protection of the catchment areas of water bodies and the marine life in them–provides an active and positive approach to the conservation of nature.

Unusual landscapes

Traditional Berekum people ascribe sacredness to landscapes that are unusual or extraordinary. This is due to their belief that the very nature of certain natural objects is hierophanic–i.e. capable of attracting spirits or revealing the divine. This has resulted in a kind of existential connection between the local people and the sacred landscapes, which promotes the conservation and preservation of these landscapes. The table below lists a few such landscapes in the traditional area.

Table 10. List of some the revered landscapes in Berekum Traditional Area

Type of sacred species	Its Nature	Examples/Location
Cave	A long and narrow cave	At *Kuum* near Koraso-Nkantanka
Odomankoma nsa (God's hand)	A stone with five fingers like those of a human being inscribed on it	In between Nanasuano and Ayimom
A stone	A huge strange stone	In the *Adeɛmmra* sacred grove at Senase
Boɔ Kofi (stone Kofi)	A huge stone	At Adesu near Kato
Bokyɛm (stone *Kyɛm*)	A huge flat stone	Near Biadan

Author's fieldwork notes

These landscapes are potential tourist destinations that could be developed to enhance the socio-economic development of the area.

The Institution of *Nkyida* or the *Nnabɔne* (Taboo days)

The institution of taboo days among the traditional Berekum people also has a bearing on the management of natural resources. The words *nkyida* and *dabɔne* (*nnabɔne* [pl.]) are used interchangeably in the Berekum community to denote what is now translated by anthropologists as 'taboo days'. *Nkyida* is a combination of two Akan words–*kyiri,* (to hate) and *da* (day). Thus, '*nkyida*' literally means 'hateful day'. Similarly, *dabɔne* comes from the two Akan words, *da* (day) and *bɔne* (bad or evil). However *nkyida* or *nnabɔne* are viewed as sacred days for the gods and ancestors of the various communities in the traditional area, despite their odd-sounding names, and hence no work is permitted on such days (except in emergencies, such as going to one's farm to get food for domestic use). There are countless stories about people who have suffered various misfortunes (and even death) for violating these religio-cultural injunctions. According to my informants, although *nkyida* or *nnabɔne* are sacred days on which cultic attention is given to objects of worship, they are viewed as hateful or bad days (hence their names) because of the harm that one can suffer for any infringement on them, as well as because of the restrictions that are placed on people's freedom on these days (*Ɔkyeame* Yankyera, personal communication, 14 October 2011, Jinijini; see also McCaskie, 2002: 153). The reckoning of these *nkyida* or *dabɔne /nnabɔne* is based on the Akan calendar. This calendar is based on what is known as *adaduanan–da* meaning 'day', *aduanan* meaning 'forty'–i.e. the forty-day cycle.

Although the cycle is known as *adaduanan*, a close examination of it shows that it contains forty-two different days, with the forty-third representing the start of a new cycle (Bartle, 1978). There are four special days within every *adaduanan*, and it is these special days that the traditional Berekum people refer to as *nkyida* or *dabɔne /nnabɔne*. Two of these *nkyida* or *nnabɔne* are called *adae* (literally translated sleeping place), and obviously referring to the sleeping places of the ancestors, since *adae* is connected with the ancestors.[1] Thus, on *adae* days, the ancestors are bathed, anointed, fed and clothed by the reigning Chiefs. That is, the chiefs visit the stool rooms[2] to: pour libation and to clean/wash the stools (bathing); slaughter either a sheep or a fowl and allow the blood to drain on the stools (anointing); use the carcass to prepare food for the ancestors (feeding them); and then re-cloth the stools (clothing them) and place them at their respective positions until the next *adae* day (*Nana* Kwasi Agyemang, personal communication, 5 October 2011, Berekum). It is important to note that the stools are placed in order of the chiefs' seniority. That is, the stool of the first chief of the community comes first, then the second one, in that order. The reigning chief's stool goes to the stool room only after his/her death (a ritual that was confirmed by all the chiefs that were interviewed).

In addition to refraining from working on the farm, no funerals are permitted on *adae* days either. The other two *nkyida* or *nnabɔne* are *fɔdwo* (on Mondays) and *fɔfie* (on Fridays). In addition to these four special days, both the state gods and those of the various communities also have sacred days, and these should be observed as holidays for those within the relevant communities (*Nana* Kwasi Agyemang, personal communication, 5 October 2011, Berekum). Certain rivers and lands may also have sacred

---

[1] For details on the Akan calendar, see Rattray, 1923: 115; Bartle, 1978: 80-4; McCaskie, 2002: 152.

[2] Among the Akan, every chief is supposed to have a stool, known as *asɛsɛdwa,* which is usually made from *sɛsɛdua* (*Funtumia* sp.). It is this stool that is blackened and placed in a special room called *nkondwafieso* ['stool room'] after the death of a chief. Therefore, all the important rituals connected with the ancestors centre around these stools. In fact, the stool is the symbol of a chief's office. Any chief whose stool is not blackened and placed among those of his or her predecessors is not a recognised chief among the Akan (Nana Kwabena Wusu, personal communication, 12 October 2011, at Nana Takyiwaa's palace in Senase).

days that prevent those who farm near those rivers or lands from working on those areas during the sacred days. For instance, no farming activity is permitted on Fridays for those who farm around the catchment area of *Asuo Koraa* in Berekum.

These restrictions placed on the use of natural resources by *nkyida* or *nnabɔne*, although potentially inconvenient and difficult for the Berekum people, nonetheless help to promote the sustainable use of the natural environment. In the study area, as many as nine *nnabɔne* could occur within one *adaduanan* cycle. Resting for nine days within an *adaduanan* cycle, apart from being good for people's health, can also give the land and other parts of the natural environment some respite to regenerate itself. It is in this light that one can see the ecological value of the institution of *nkyida* or *nnabɔne* for the people of the study area, and throughout Ghana, as *nkyida* or *nnabɔne* are observed in every traditional society within Ghana. Thus, it can be argued that the institution of *nkyida* may have some scientific underpinning, since it provides nature with a respite to regenerate itself.

Planting *Ngyedua*

The practice of tree planting in Berekum Traditional Area dates back to as long as the community itself. It is a common sight to see species of trees planted in the principal streets of all the towns and villages in the traditional area in order to divide the streets and improve their aesthetics. Also, every important occasion in towns and villages climaxes with the planting of a tree in order to help with the reckoning of dates in Berekum society. Such trees are referred to as *gyedua* (*ngyedua* [pl.]).

Figure 17 The researcher in a focus group discussion under a gyedua at Fetentaa

The trees provide good places for the youth to meet and enjoy fresh air, as well as to play games such as draughts, *oware* and others. Even I had the privilege of conducting one of my focus group discussions under a *gyedua* plant at Fetentaa village (see fig. 17). At times, settlements of disputes take place under these trees.

Although planting trees in the towns and villages used to be very common in Berekum, many of these trees have been felled to make way for the erection of electricity poles (*Ɔkyeame* Okra, personal communication, 7 November 2011, at his Biadan). Whatever the case may be, planting *ngyedua* is a traditional strategy which can be used to address the problem of environmental degradation in Berekum and in Ghana more general. Furthermore–and more importantly–the traditional Berekum and the Akan people believe that certain plant species have the ability to ward off evil spirits. Therefore, plants serving as the *genius loci* can be found in every Akan village as the founding ancestors of communities would plant such trees there for this purpose (*Ɔpanyin* Yaw Mensa, an 80-year old man, personal communication, 10 October 2011, Fetentaa).[3] A ritual known in Akan as *gyinae* normally accompanies the planting of such trees. This ritual is seen as a foundation for beginning a settlement (a town or village) among the Akan, and has preceded the beginning of every Akan settlement–a fact attested to by all my key informants. Trees such as the *Nyamedua* (God's tree), *Sinuro* (*Alstonia gongenis*), *Sumɛ* (*Costus afer*), and the fig tree are usually used for such purposes.

Unlike those areas in which trees are used for beautification purposes, people are usually restricted from entering places where *ngyedua* are used for *genius loci* purposes. Many *ngyedua* are large-leafed trees of the ficus family. A common one in Berekum is known as *amangyedua* (literally translated as state tree). Since such trees serve as *genius loci* for communities, the community members (especially the gerontocracy) take proper care of them as the death of such trees is seen as a bad omen for the entire community. Hence the places where such trees stand also become shrines for the entire community.

The ecological importance of these indigenous practices may be seen in the fact that some species of plants are planted to serve as windbreaks for protecting the environment. Thus one can argue that, in addition to the religious underpinnings of indigenous ecological practices, some of them also have practical and scientific underpinnings.

Belief in *Sasa*

Another basis for indigenous ecological practices that have a religio-cultural underpinning is the belief in *sasa--tumi* (spirit) believed to be living in certain plants and animals (see Awuah-Nyamekye, 2012a: 4;

[3] Parrinder (1974: 52) and Rattray (1923) also described this practice in their studies.

2009a:.269). The traditional Berekum people, like other Akan and traditional Ghanaians, perceive certain species of flora and fauna to be sacred, believing them to possess *tumi* (spiritual power). *Sasa* is the Akan word used to denote the kind of *tumi* suffused in those flora and fauna (*Ɔpanyin* Kwasi Broni, an experienced hunter, personal communication, 8 November 2011, Senase; see also Awuah-Nyamekye, 2012a; Rattray, 1927; 1959). In the traditional Berekum worldview, the created world is composed of countless individual beings and objects that nonetheless exist in a unitary form due to their possession of *sunsum* (spiritual force), which is believed to ultimately derive from the *Onyame* (Supreme Being), the Creator, sustainer and source of all existence. Those individual beings with higher *sunsum* are termed as *sasa* (*Ɔpanyin* Kwasi Broni, personal communication, 8 November 2011, Senase). An important feature of *sasa* is that not all spirits are of equal standing, with some *sasa* seen as being more powerful than others. Hence, there are *sasa a ɛyɛ hare* (*sasa* which are light or not heavy) and *sasa a ɛyɛ duru* (*sasa* which are heavy) (*Ɔpanyin* Kwasi Broni, (Personal Communication, 8 November 2011, Senase; see also Awuah-Nyamekye, 2012a; McCaskie, 2002).

Those trees and animals that are believed to have light *sasa* are perceived to be harmless to humans, but those with heavy *sasa* are perceived to be dangerous, as they are seen as having the capacity to be vindictive (this is a widely held belief among the Akan–see McCaskie, 2002). As such, the people tread cautiously with such plants and animals. In the study area, the following plants were mentioned as some of those that have heavy *sasa* suffused in them*:* *Odum* (*Chlorophora excelsa*), *Esa* (*Celtis mildbraedii*), *Abeko* (*Tieghemella heckelii*), *Tweneboa* (*Entant drophragma), Onyina* (*Ceiba pentandra*), *Homakyɛm* (Dalbergia saxatalis), *Odii* (*Okuobaka aubrevillei*). The animals include: *Trɔmo* (*Tragelaphus euryceros isaacii*), *Oyuo* (*cephalophus niger*), *Kuntun or Pataku*–striped jackal (*Canis adustus*), and *Ɔkoɔ* (buffalo) (see also Awuah-Nyamekye, 2012a: 4). Sarpong (1974) notes that plants with *sasa* are seen as creating the 'residential areas' for spirits and gods, and thus they must not be felled without the appropriate rituals being performed.

It is not only some of the animals and plants that are believed to possess *sasa* in the study area, but some of the human beings as well. However, in human beings, the *sasa* are believed to be potent only when the individual is dead, and they are believed to be especially potent in people who die in violent ways or who have had unbefitting burial rites performed on them. However, the issue of *sasa* in human beings is outside the purview of this book.

To see why Berekum people (and traditional Ghanaians in general) are cautious about entities they believe to have *sasa*, I will now provide two examples to illustrate how this belief operates in the lives of the traditional Berekum people. One plant believed to possess *sasa* is the *Odii/due* (*Okuobaka aubrevillei*)–one of the most respected and feared plants among the traditional Berekum people and other Akan. *Odii* is considered to be the 'king' of all plants in traditional Akan societies (*Ɔpanyin* Kwame Anane, a herbalist, personal communication, 7 October 2011at Mpatasie, see also Awuah-Nyamekye, 2012a).

Many of my informants said that it has no fallen leaves under it, and it was explained that only very few animals can pass under it–any animal that is not spiritually powerful will die instantly if it passes underneath. The hunters I interviewed claimed that they see many bones of dead animals under the tree. They said it is only the *Okusie* (Giant rat) and the *Apɛsɛɛ* (*Artheerus africanus*) that could pass under the *Odii* without any problems. This implies that those animals are believed to have strong spiritual powers. Paradoxically, however, these two animals are those that are often the most hunted in the Berekum area. Unfortunately, my informants could not explain why this is so. It may be due to the palatable nature of the meat of these two animals or it may be due to, as Cox (2000) and others have commented, the rate at which local knowledge systems are disappearing, we may not even know what purpose, if any, such practices have (Brodt, 2001; Pandey, 2003). In the study area, medicine men and women use the bark of the *Odii* to treat various ailments, including skin diseases and venereal ones like syphilis, but it is commonly believed that only spiritually powerful traditional medical practitioners can produce medication from them.

Another plant that it is believed to possess *sasa*, and which thus commands awesome reverence, is the *Homakyɛm* (*Spiropetalum heterophyllum*). This plant is a type of liana. My informants claim that the sap in this plant is the colour of blood, and the hunters I interviewed claimed that this plant could talk like a human being during the night. Medicine men and women in the study area believe that *Homakyɛm* has a very high medicinal value, but say that only traditional medical practitioners that are spiritually fortified can cut this plant for medication. It was said that even those who are capable of using parts of this plant for medication have to do so either early in the morning or late in the evening, because if one's shadow falls on it whilst one is cutting it, one will be spiritually cutting oneself. Therefore, this plant has to be cut when there is no sunshine to cast a shadow. The medicine men and women added that because of the fear attached to this plant; those who cut it have to propitiate it before doing so (see also Awuah-Nyamekye, 2012a:.5). Research that McLeod (1981) and Falconer (1992) carried out among Akan communities supports this finding. *Nkosua* (eggs) and *nsa* (drinks–generally alcohol) are usually used in accompaniment with prayer as part of this propitiating ritual (see also Awuah-Nyamekye, 2012a:.5).

Animals believed to possess *sasa*

Among the people of Berekum Traditional Area, the *trɔmo*–the bongo antelope–is the animal most feared by hunters due to the dangerous *sasa* that it is believed to possess--a claim made by all the hunters I interviewed. McCaskie (2002: 220) and Rattray (1959: 182) generated similar findings among the Ashanti. The *sasa* of this animal is believed to be more powerful than the *sasa* of any other animal in the animal kingdom, and it is held that when it is killed, its *sasa* can haunt the hunter until it is exorcised through a special ritual bath. The hunters that I interviewed were unanimous on this matter. It was said that when one is affected by the *sasa* of a *trɔmo*, one suffers from hallucinations, through which one will keep shouting the animal's name–'*trɔmo! trɔmo! trɔmo*!' The hunters claimed that an affected hunter could even mistake a human being for an animal and shoot at them. They said, this could lead one to make false confessions (see Awuah-Nyamekye, 2012a: 7).

Figure 18 The trɔmo (the bongo)
Source: http://en.wikipedia.org/wiki/Bongo

There is a popular saying among the Akan that runs: '*Yɛnni sasa aduro a yɛnnkum trɔmo'*, which translates as 'If you do not have *sasa* medicine you do not kill the bongo' (see Agyemang 199:.144). Another saying runs: '*Sɛ wobɛkum trɔmo na woadware sasaduru deɛ, gyae no ma mfa ne mmerantesɛm nante kwae ase'*, which translates as 'If you kill the bongo its spirit will haunt you until you exorcise it; you better leave it to roam about in the forest'. Because of the mystery surrounding the *trɔmo*, many hunters in the area avoid it altogether during their hunting expeditions

Another animal that hunters of the study area avoid is the *Kuntunu or Pataku*–the side-stripped jackal (*Canis adustus*) (Awuah-Nyamekye, 2012a)–again, because of the dangerous *sasa* it is believed to have. It is even said that its presence can bring about confusion in the community. My informants hinted that a sorcerer can cause a serious fight to break out in a community for no justifiable reason by just shaking the tail of a dead side-stripped jackal (*woso Kuntunu dua*). For this reason, hunters are hesitant to hunt this animal (Awuah-Nyamekye, 2012a:.7). In the study area, many other animals, such as elephants, leopards, pythons, buffaloes, crocodiles, and black duikers, must have their *sasa* propitiated when they are killed to prevent them haunting the killer (*Ɔbɔfoɔ* Yaw Barima, personal communication, 30 August 2011, Mpatasie; see also Rattray, 1959: 183). Parrinder's (1974) study across Africa confirms these findings. Thus, it may be argued that the belief in *sasa* has the potential to contribute to the protection, conservation and sustainable use of certain trees and animals in both the study area and Ghana as a whole.

Figure 19 . The side-stripped jackal
Source: http://en.wikipedia.org/wiki/File:Side striped_Jackal.jpg

It is important to state that because the above discussed traditional ecological measures have religious undertones, the infringement of any of them is viewed as a taboo or a 'sin' in the life and thought of the people under investigation. The measures can also be viewed as traditional or indigenous ecological laws. This is because anyone who infringes any of them in spite of the belief that one may suffer the consequences spiritually, one is also punished by the leadership (Chief and elders) of the community, which in most cases is in the form of payment of a fine and/or provision of specified items for sacrifice to appease the ancestors or the spirit entity that is involved. This means or confirms the view that in Traditional Berekum Society and as it is the case in other traditional

African societies, religion and politics are mixed (Odotei, Awedoba and Albert. 2006; Owusu Brempong, 2006, Awuah-Nyamekye, 2009).

As pointed out in the introduction, some literature take issue with parts of the religio-cultural practices above. For instance, critics of religious environmentalism argue that there are numerous examples of the (so-called) nature religions (including African Traditional Religion) that only concentrate on certain aspects of nature, such as trees, forest or rivers, in contrast to the focus of modern environmentalists, who attend to the entire environment (see Tomalin, 2013). In the words of Taringa (2006), 'this means not all aspects of nature play a pivotal and vital role in their beliefs about salvation … [since] some aspects are treated with [the] least care and reverence' (2006: 211). Although my interviews revealed that the people of Berekum Traditional Society place more emphasis on trees, forests and rivers than on other elements of creation, they also revealed that they perceive the whole of nature to be interconnected, and thus requires human attention to ensure that harmony is maintained. Although they take human subsistence needs to depend more on certain elements of nature than others, and thus exploit some of these aspects of nature for their own survival, they do so only in a way that does not pose an overall danger to nature. Thus, they interact with nature in a sustainable way. *Ɔpanin* Kyere Kwame's and Nana Akumfi Ameyaw's statements in Chapter Four address some of the concerns of the critics. A repetition of Nana Akumfi Ameyaw's explanation here may be relevant:

> In God's creation, there are things that can be used (destroyed) and others that cannot be used, so the destruction of such things constitutes a sin. Even those that can be used have the right time for their use. Anything short of this is considered as a sin in our tradition. This is to ensure that we make sustainable use of natural resources (Personal Communication, 7 November 2011).

Some of the Berekum proverbs that were presented in chapter five also confirm that the critiques of religious environmentalism discussed above hold little sway in relation to the people under study, who clearly understand what goes into their environmental ethics. In view of this, it is contended that the arguments put forth by critics of religious environmentalism do not find purchase in all indigenous societies.

In spite of the above explanation, it will be misleading to deny the ambivalent way in which indigenous Africans relate towards nature. In other words, some of the religio-cultural practices may be seen as injurious to the environment. That is, aspects of ambivalence are embedded in indigenous Africans' religio-cultural practices which raise issues with the view that indigenous African religion is 'environmentally friendly' (Schoffeleers, 1978). This in a way makes such a claim idealised or romanticised. For instance, among the people under investigation and indeed throughout Ghana, there is less respect towards non-totemic animals or plants. The only non-totemic animals or plants that are treated with awe are those, as has been pointed out earlier, believed to possess *sasa*. This seems to confirm Taringa's (2006) view that environmental friendliness of indigenous Africans is selective. Also among the Effutus of Ghana, their major festival is *Aboakyerɛ* (lit. animal catching) during which the deer is hunted for as part of the major rituals of the festival. This indeed puts the life of deer in the community in danger. This has begun to show as during last year's (2012) festival no deer was caught in spite of persistent search. Furthermore; the *Bugum Chugu* (Fire Festival) of the Dagomba people of northern Ghana is a potential source of bush fire. This is because one of the main rituals of the festival is where the participants gather usually in a spot at the outskirts of the town with everybody holding high his or her long burning torch. These torches are made of dry grasses specifically prepared for the occasion. The dry nature of the grasses makes it more serious since the vegetation of the place is overwhelmingly grassland and usually dry due to the arid weather condition in the northern part of Ghana.

Summary

The forgoing discussion has spelt out how the Berekum people have been influenced by their indigenous religion and culture in relation to the environmental conservation. These people hold a firm belief that human beings live in a relationship with other creatures of the universe, and thus need to establish an acceptable way to sustain this relationship for their own benefit. Armed with such a belief, they have used mechanisms such as sacred groves, *nkyida,* taboos, reverence for land, water bodies, the belief in totems and *sasa*, and the planting of *ngyedua* to help them to relate positively with the environment with a view to conserving it.

Although the origins of sacred groves–an important traditional conservation strategy–and the reasons for their establishment appear to be religious, a deeper reflection reveals the presence of conservationist motives. Whatever it is, it I observed that sacred groves do have some ecological importance, as other studies have confirmed. For instance, notwithstanding their size, they are found to serve as sanctuaries for the wildlife inhabiting them.

Both the fear and the respect for certain plants and animals grounded in the belief in *sasa* and the practice of totemism, respectively, also serve as concrete ways for conserving a variety of species. In much the same way, reverence for land and water bodies, as well as the practice of planting trees, was identified to have positive ecological values in the study area. In spite of these positive religio-cultural practices that enhance the weal of the environment, there are other religio-cultural practices that do not support environmental conservation. In the next chapter, the means by which Berekum people create and transmit indigenous ecological knowledge is examined.

CHAPTER SIX

INDIGENOUS BEREKUM WAYS OF CREATING AND TRANSMITTING ENVIRONMENTAL KNOWLEDGE[1]

Introduction

> You cannot protect the environment unless you empower people, you inform them, and you help them understand that these resources are their own, that they must protect them. (Wangari Maathai, 2011)

The above quote by the prominent, late Kenyan environmentalist, Wangari Maathai shows the importance of education if environmental conservation programmes are to succeed. For this reason, the people of Berekum Traditional Area have put some measures in place to ensure the sustenance of their indigenous ecological knowledge. Understanding the value of education that imparts environmental knowledge and behaviour, traditional Berekum people use different methods to pass on their ecological knowledge to the youth. This section examines some of these methods, of which there are several. The most important ones are discussed below.

Mmɛbuo (Proverbs)

The Akan word for proverb is *ɛbɛ* [*mmɛ* [pl.], whilst the practice of using proverbs is referred to as *mmɛbuo*. *Mmɛ* are used in context, and in order to use them effectively, one has to be knowledgeable about Berekum culture, particularly the language of the community. Corollary, the listener has to also be well versed in the culture, particularly in the idioms of his

---

[1] This chapter is adopted (with permission) with some modifications from my previous publication in the *Journal for the study of Religion, Nature and Culture*, 8 (1):46-63, 2014.

society so that he/she can appreciate the importance of the message being conveyed in the proverb(s). Consequently, traditional Berekum people make sure that their youth become proficient in the use of the idioms of their community. Makilah describes a proverb as 'a saying that expresses an experience that is considered an apparent truth within a particular community' (Makilah, 2003: 1; cited in Omare, 2010: 90). Because the youth are prepared to appreciate the use of *mmɛ,* the use of *mmɛ* provides a common tool for imparting knowledge to them. *Mmɛ* are used to create an awareness of other creatures through reference to their natures and particular characteristics. For instance, the word *ako/awidie* (parrot) is used to emphasise oration or eloquence. It is common to hear expressions such as '*na'no ate sɛ ako/awidie*', meaning 'he/she is as eloquent as the parrot'. According to Nana Kwabena Wusu (Personal Communication, 7 November 2011), some of the *mmɛ* advise people to be wary of certain plants and animals due to the magical powers or evil spirits that are believed to be associated with them. For instance, the following two proverbs (which I have already mentioned) are commonly used in the study area: '*Yɛnni sasa aduro a yɛnnkum trɔmo*–'If you do not have *sasa* medicine you do not kill the bongo'; and '*sɛ wobɛkum trɔmo na w`adware sasaduru deɛ, gyae ma ɔnfa nemmrantesɛm nante kwae ase*'–'If you kill the bongo its spirit will haunt you until you exorcise it; you better leave it to roam about in the forest'.

These *mmɛ* are used to warn people to be wary of the bongo. Because Berekum people are fully aware that the continuous survival of humans is possible only if natural resources continue to be available to them, the *mmɛ* are sometimes used to draw attention to the sustainable use of resources. This explains why it is common to hear a proverb such as 'a*didaa yɛ sen adi prɛko*'–'it is not good to eat all that you have in a day', or '*wosum boɔdeɛ a sum kwadu*'–'when you support the plantain plant against the wind, do the same for the banana plant'. The contents of the *mmɛ* are usually compounded with those of personal experiences, observations, and information about the reality of natural phenomena. That is, every *ɛbɛ* in Berekum is meant to teach a lesson or draw attention to something. Therefore, all the *mmɛ* that say something about natural objects create awareness about their existence. In other words, through the use of *mmɛ* people, particularly the youth, are equipped with knowledge of the physical environment including the animals of the jungle. It is, therefore, not surprising that a lot of *mmɛ* abound in this traditional society, and provide prescriptions about how humans should relate with natural phenomena.

The following table provides some examples of common *mmɛ* in the study area that relate to the human-nature relationship. It was compiled during the interview sessions with my key informants.

Table 11. Samples of human-nature proverbs in Berekum Traditional Area

Proverb	Literal translation	Kind of relationship involved	Meaning/ Relevance
Nnua nyinaa tutu a ɛbɛ ka abɛ.	It is only the palm tree that is able to withstand the strength of a destructive storm.	Plant-plant	Dependability
ɛnam dua so na ɛma homa hunu soro.	It was through the help of the tree that the vine was able to reach the sky.	Plant-plant	Cooperation
Ayɔnko gorɔ nti na ama ɔkɔtɔ ɛnya tiri.	It was having too many friends that caused the crab its head.	Animal-animal	Carefulness in friendship
Abosua te sɛ kwae wogyina akire a ɛbomu koro wobɛn ho a, na wo hunu sɛ dua bi ara wo ne sibrɛ.	Family is like a forest; when viewed from afar it looks as if all the trees are one, but when you get close you will realise that each tree stands alone.	Human-plant	Cooperation, but self dependence is important
ɔbonto nfa woho mmɔ nkaawa.	*ɔbonto* [a kind of fish] do not compare yourself with *nkaa* [another fish, believed to be very cunning].	Animal-animal	Be selective in friendship
Dua bata boɔ netwa yɛ twa na.	It takes the wise person to cut a tree that is attached to a stone.	Plant-stone	Perseverance and carefulness
Kwae a agye wo no yenfrɛ no kwaewa.	Do not be ungrateful to the forest that has saved you.	Human-plant	Gratefulness
Dua koro gye mframa a ɛbu.	An isolated tree cannot stand the might of a strong wind or storm.	Plant-wind	Cooperation
Aserewa su ayankuku su a neto pae.	When *Aserewa* [a kind of tiny bird] tries to imitate the cry of *ayankuku* [a bigger bird] its anus bursts.	Animal-animal	Be wary of over ambition
Sɛ wo bribi ne abob ɔnnua a ɛnka wo bɛtɔ akuma ama no	You would have bought an axe for the woodpecker if it had been one of your relations.	Human-animal	Sympathy
Aboa a ɔnni dua Onyame na ɛpra ne ho.	It's God that drives away flies for the tailless animal.	Animal-God	Concerned/ Care for the needy

Author's fieldwork note

What may be drawn from the table above is that the contents of many Berekum proverbs are derived from human-plant/animal relationships, plant-plant relationships, and animal-animal relationships. This implies that the contents of proverbs help to create or emphasise the idea that human existence depends on the existence of non-human creatures. Asante-Darko's (2006) study of the relationship between Asante proverbs, poetry and environmental knowledge produces similar findings. For instance, the proverb '*kwae a agye wo no yɛnfrɛ no kwaewa*'–which literally means 'do not be ungrateful to the forest that has saved you'– covers a wide range of issues. Wienecke's (2005: 88) study also recognises the positive value that proverbs can have in inculcating the need to conserve natural resources.

Socially, proverbs are used to impart a sense of gratitude in the youth. Philosophically, they advise against showing ungratefulness to any phenomena that one's life depends on. The use of the word 'forest' in the above example reinforces its ecological importance, as livelihoods depend on the land (forest) in traditional societies. Thus this proverb is clearly cautioning both the young and the old to be grateful to the forest as it is an important natural resource, and as such needs to be used sustainably.

The proverb, *'dua koro gye mframa a ɛbu'*–which literally translates as 'an isolated tree cannot stand the might of a strong wind or storm'–is used to stress the social need for cooperation in all aspects of human life. The need for cooperation in the worldview of the people under study is not limited to human beings, but is also important for other constituents of creation. This view is expressed in another proverb, '*dua koro nyɛ kwae*' ('one tree does not make a forest'). The ecological lesson these two proverbs emphasise is that a lot of trees are needed in the same place if you want to get the full benefits from them. Each of the proverbs in the table above will, in one way or another, emphasise the fact that human survival is tied to the rest of nature, and particularly to trees. Thus, all Berekum proverbs that make reference to elements of nature can be used to impart ecological knowledge.

The awareness that proverbs create about nature has accordingly influenced the traditional Berekum people's attitude towards the non-human creatures that play a meaningful role in their lives. It should be noted that I am not suggesting here that environmental anthropocentrism (valuing nature solely for its benefits to humans)–an approach that deep ecologists fervently oppose (Sessions 1995, Naess 1973)–is the main driving force for Berekum people's attitude towards nature. The proverbs exist in this traditional milieu to show that they value nature through a holistic understanding of it, not because they are purely instrumental tools for promoting the self-interest of Berekum society. For instance, traditional Berekum people believe that everything God created is good. They express this proverbially as: '*Asansa se nea Onyame abɔ nyinaa yɛ'*– 'the hawk says all that God has created is good'–which is a commonly used proverb among all the Akan of Ghana (Opoku, 1978).

Just as Kothari (2007: 4) rightly points out, 'some forms of traditional knowledge are expressed through [myths] stories, legends, folklore, rituals, songs, art, and even laws'

Myths

Within the Berekum traditional worldview, myths are classified under the broad term *anansesɛm*2 (Spider's stories). But myths are '*anansesɛm*'with a difference. This is because *anansesɛm* are ordinary stories told to entertain, educate or to teach a particular lesson, whilst myths are generally believed to be true stories in the community in which they are told. Myths offer the "why's" and "how's" of the existence of religio-cultural beliefs and practices, as well as of other important elements in a community. However, in an interview with Nana Agyemang (Personal Communication, 6 October 2011), he said that there is no specific local term for what we now refer to as myths, and that is why they are loosely classified under the term *anansesɛm.* He added that myths, unlike *anansesɛm*, are not ordinary stories, and this may explain why only a few people (mostly the elderly and ritual specialists) are capable of narrating the myths of a particular community.

The above explanation corroborates the literature on myths, where there is no unanimity over their definition among folklorists (Bascom, 1965). Bascom for instance, defined myths as 'prose narratives which, in the society in which they are told, are considered to be truthful accounts of what happened in the remote past' (Bascom, 1965: 4). He explained further that myths give account of the origin of the universe, humankind, death, the nature of some animals, or of any natural phenomena, and may explain the origin of a ritual or a taboo (Bascom, 1965: 4), while the *Encyclopaedia Britannica* explains that a myth 'is a symbolic narrative, usually of unknown origin and at least partly traditional, that ostensibly relates actual events and that is especially associated with religious belief.'

---

[2] The spider is one of the most important characters in Akan folklore and West African folklore more generally. In the Akan and traditional Berekum mythology, *Ananse's* character is ambivalent, standing for both wisdom and trickery because of its ability to adjust craftily to situations (see Rattray, 1930). For that reasons, Akan, folklores are called *anansesɛm* (i.e. they are named after the spider).

Thus, myths are usually used by indigenous people to explain the beginning of events, and traditional Berekum people are able to use them to educate the youth about the rituals and origins associated with the beliefs and practices of the area. For instance, there is a myth among the people of Berekum that explains why fish and crabs in the River Koraa are not eaten. Maame Akua Donkor (the attendant of the *Koraa* sacred grove, personal communication, 7 March 2009) recounts how 'in the olden days', a serious war broke out between the Berekum people and their neighbours, the Dormaa people. But the River *Koraa* miraculously saved the Berekum people by overflowing its banks whenever the opposing army tried to launch an attack on them. The people of Berekum considered this a miracle because it was during the dry season that the river's banks overflowed. She said that prior to the war, the people could eat fish and crabs from the river, but immediately after the river saved them, they began to experience stomach upsets whenever they ate either a crab or a fish from the river. An oracular consultation revealed that the river did not want her 'children' (fish and crabs) to be destroyed. This explains why fishing is strictly prohibited in the River *Koraa*. The people now go to the riverside to watch the fish as they swim beautifully in the stream

The above story shows how myths can be used to explain why certain ecological practices exist in a particular community. LEAF (2009) cites Arnes Naess as advising against viewing scientific knowledge as being completely incompatible with myths, recommending: 'Let us value both the myths and the scientific studies in order to develop clearer understandings about our precious forests'. Myths, by their nature, can enable ecological knowledge to be passed on to younger generations through explanations of the origins of human beings, the cosmos and its constituents, together with the provision of guidance about how humans should relate with these features of the universe. Herein lies the importance of myths in generating and transmitting ecological knowledge, particularly in indigenous societies.

Just like myths, folktales are used by the traditional Berekum people and the Akan in general to explain the existence and origins of natural phenomena, thereby giving an account of human beings' relationships with them, and it is to these cultural devices for transmitting knowledge that I now turn.

Folktales

Discussions with my key informants revealed that folktales are also classified as *anansesɛm.* Folktales are stories that elders in the community use to impart knowledge and entertain younger members of the community, usually during the evening (see also also Mawere 2014; Mawere, M. 2012). In some of the folktales, elders are able to create an awareness of natural phenomena and human beings' relationships with them. The following folktale attests to this:

> Once upon a time, there existed a close friendship among three animals– *Tatia* [ant], *ɔkra* [cat] and *ɔsono Kokroko* [the almighty elephant]. The friendship grew in strength. One day, the friends decided to discuss how each would honour his mother when she died. *Tatia* was first to speak. He told his friends that his mother was so dear to him that when she died, he would give her a special honour by laying her on a skin of a human being. *ɔkra* also said he would kill a human being to honour his mother when she died. *ɔsono* said, when the mother died, he would honour her by destroying all the trees around him to show that he has the strength and power over everything in the world. Sooner than later, *Tatia* lost his mother. He was able to honour his promise by indeed laying his mother in 'state' on a skin of a human being. All the mourners became surprised at this feat. *Tatia* was asked to explain how he managed to get the skin of a human being. He explained that he got it from the peeling offs of the skin of human. He explained further that any time a human being gets a wound or a burn, and it is getting healed, part of the skin where the wound or the burn occurred peels off. He said that it was these peel offs that he laboriously gleaned over the years. It was from the piles that he made the 'mat' (human skin) on which his mother's body was laid in state. *Tatia* was applauded for his tenacity, ingenuity and dexterity by all the friends and sympathisers at his mother's funeral. This also explains why it has become the character of the ant to always move slowly picking items from the ground. *ɔkra*'s mother was second to die but he was unable to kill a human being in honour of his mother as promised. He was booed at for failing his mother. Out of disgrace, *ɔkra* decided to make amends by promising to rub his body against the body of any human being he comes across. This explains why *ɔkra* is fond of rubbing his body against humans' bodies all the time. When *ɔsono Kokroko*'s mother died, he started destroying trees around him as promised, but he could not pull down the *odum* [*Chlorophora excelsa*] tree. He tried several times but to no avail. He became disappointed and ashamed. He also was booed at by his friends and sympathisers at the funeral for failing to honour his promise to his mother. He was told that it is honourable to be modest in one's promise and that one should not be deceived by his or her mere size to claim to be invincible (Stephen Effah Mensah, personal communication, 27 January 2012).

Although there is a strong emphasis on social lessons in the above story, an ecological lesson is also implied here, as the protagonists of the story are animals dealing with human beings and flora. In the words of Asante

(2005), 'the animal and plant characters in folktales are presented to allow the narrator and audience to stand back and contemplate the fact that they have lived in close contact with them' (Asante, 2005: 106). In other words, through the use of 'characters of both animate and inanimate things interacting, the audience is being reminded of all the biodiversity that is outside them yet forms part of this arena of their world and constitutes a basis of their lives and sustenance' (ibid.). Furthermore, Mbiti (1966: 23) suggests that the use of non-human creatures in folktales implies that 'they [non-human creatures] contain a portion of humanity just as humanity contains a portion of them, and none is independent of the other'. Also, the very presence of flora and fauna in the story facilitates the smooth impartation of socio-moral lessons to the members of the community, implying that flora and fauna are necessary ingredients for human existence and survival, and thus placing a perceived obligation on humans to protect them. Finally, the idea of recycling could be promoted by the ingenuity of the ant.

In addition to such types of stories, Berekum adults use everyday examples to educate their young about environmental issues. For instance, when dealing with a difficult or dicey situation, one will sometimes hear an elder from Berekum say, '*Ayɛ sɛ santrofi anoma. Sɛ wokum no a, woafa mmusuo; sɛ wogyae no nso a, wagyae seradeɛ*', which literally translates, 'the situation is like the nightjar [a nocturnal bird], which, when you kill it, you invite misfortunes on yourself but, when you leave it, you have missed a delicacy'. Although this saying is intended to advise people to be very careful when taking a decision, it also has the potential to warn people to avoid this kind of bird.

Bragorɔ (Puberty Rites)

Certain rituals are also used in Berekum culture to impart ecological knowledge to young people. Three of these rituals–*Bragorɔ, Afahyɛ* and *Yɛrepra* yadeɛ–are discussed here.

Bragorɔ[3] is an important event and occasion for teenage girls in Berekum Traditional Area as it is a formal means for integrating teenage girls into the community. The rites for this ceremony are performed as soon as a girl experiences her first menses (Nana Adwoa Takyiwaa,

[3] For details on Akan *bragorɔ* see Sarpong, 1977 Girls Nubility Rites in Ashanti, Accra: Ghana Publishing Cooperation. Also, see Appiah-Sekyere, P & Awuah-Nyamekye, S .2012. 'Teenage pregnancy in the life and thought of the Akan: Moral perspectives', *Sociology Study*, vol. 2, no 2,:.129-138.

Senasehemaa Queen Mother of Senase) and *Kurontihemaa* of Berekum Traditional Area, personal communication, 12 October 2011, at her palace in Senase). This may be described as being both an initiation and a separation ritual, as the girls are simultaneously being separated from the domain of childhood and introduced into adulthood. Thus, it is a formal way for inducting girls in Berekum into the customs and traditions of their people.

During the ceremony, the girls are secluded for a period of time, during which they are educated in matters of sex, marital and family matters, and the taboos of the land, especially the environmental ones. Nana Takyiwaa explained that although the primary importance of *bragorɔ* rites is the provision of an official platform on which a girl is declared matured for marriage in traditional Berekum society, the occasion is also used to teach the girls the "dos and don'ts" of their community in general. For instance, they will be made aware of why they should not go into rivers and streams, and why entry into certain places is restricted to women during their menstrual cycle, as well as why certain foods should not be eaten during pregnancy. Therefore, *bragorɔ* rites are meant to aid girls in their transition into the adult community. This information was corroborated by Nana Ameyaa (a ninety year-old woman) during a separate interview.

There is no formal event to induct boys in the area into adulthood, but a boy must follow his father closely and learn the customs and traditions of the land. For instance, as he follows his father, he will learn what animals and plants not to kill, either because of totem or fear of the *sasa* within them. For instance, the teenagers in the study area would be told not to kill a *pomponyin* (lizard) to avoid getting a swollen umbilicus (navel). According to *Ɔpanyin* Yaw Owusu (Personal Communication, 11 November 2011, Biadan), the only important milestone signifying the maturity of a boy in Berekum is when his father buys him a gun and a cutlass. This is significant because it means that the boy is now mature enough to take a wife. It is interesting to note that Driver (1941) has found very similar initiation rites for girls and boys entering adulthood among Native North Americans.

Afahyɛ (Festivals)

Afahyɛ are annual events that traditional African peoples celebrate. Most *Afahyɛ* are for honouring and remembering the founding fathers of the community concerned, or for honouring their gods, although they sometimes also serve to re-enact important occasions connected with the society concerned (Nana Okofo Dartey III, personal communication, 14

December 2011, at his palace). Some of these *Afahyɛ* are connected with the agricultural practices of a people, and may either mark the beginning (planting) or the end (harvesting) of an agricultural season. One such example is the *Hɔmmɔwɔ* of the Ga people of Ghana. Among the people under study, *Kwafie*–the principal festival of the people of the Berekum Traditional Area–is celebrated to honour the founding fathers of Berekum. In addition to *Kwafie*, every town or village in the traditional area has its own local festival, with its own particular purposes. I was able to compile information about some of these festivals, which is presented in the table below.

Table 12. A table of Festivals in Berekum Traditional Area

Name of Town/Village	Name of festival	Time of celebration	Purpose/Objectives
Abi	*Abofogya*	Between Nov. & Dec.	To ask for the blessings of the ancestors
Akroforo	*Afuofu*	Between Nov. & Dec.	To ask for a bumper harvest and the protection of the town
Ayimom	*Kwafie*	December	To pacify the gods and ancestors and ask for protection and a bumper harvest
Berekum	*Kwafie* (the main festival of the traditional area)	Between Nov. & Dec.	To pacify gods and the ancestors and ask for their continuing protection
Biadan	*Kwafie & Tankwasi*	December	In honour of the ancestors and the Tankwasi god, to ask for a bumper harvest and protection
Jamdede	*Koraa afahyɛ*	December	In honour of the river Koraa for protecting the community throughout the year
Jinijini	*Munukwasie*	September	To taste the 1st yam of the year and ask for a bumper harvest in the following year and the protection of the community
Koraaso	*Munufie afahyɛ*	Between Nov. & Dec.	For celebrating a bumper harvest

Kato	*Mfansi*	Every Friday	To ask the gods for the protection of the community
Kutire no. 1	*Fofie*	January	To ask the gods for a bumper harvest and the protection of the community
Kutire No.2	*Munufie*	Between Dec. & Jan.	To ask the gods for a bumper harvest and the protection of the community
Mpatapo	*Fofie*	January	In honour of the gods and ancestors, and to ask for a bumper harvest and the protection of the community
Mpatasie	*Krufie afahyɛ*	December	To remember their ancestors and solicit their protection
Senase	*Tansie*	January	In honour of the Tansie god to ask for a good yield and the protection of the community

Author's fieldwork notes

During festival periods, most of the citizens of the area return home to honour the occasion, from both near and far. As part of the celebrations, the history of the community is recounted, the dos and don'ts are re-emphasised, and those that need amendments or repeals are revised. This annual ritual reminds the people of their responsibilities and rights in the community and enables them to learn or re-familiarise themselves with the tenets of the community, including those connected with environmental conservation. More importantly, as part of the preparation for the *Kwafie*, all the roads leading to the rivers and streams are cleared, anything that is considered filthy is removed from every household, and choked drains are desilted. This is part of the cerebration of festivals throughout the traditional area, and is meant to safeguard the health of the people in the community (all the chiefs and the elders I interviewed stressed this important fact). *Kwafie* also has important social aspects. It is a period in which all the disputes among the people are supposed to be settled, and provides an opportunity for the youth to select their life partners (since all the citizens are to be present during the festive occasion). Thus, festivals have multiple functions in Ghana, and this has led the government to show an interest in some of the major ones. They accept invitations from the traditional leaders of the communities concerned as they view the festivals as good platforms for disseminating government policies and sharing national resources. It has recently become fashionable for politicians, particularly government officials, to announce new projects for communities during festival times. These officials also take the opportunity to reiterate the need to protect the environment, whilst the chiefs and their people use the occasion to make formal requests to the government for new projects. This platform serves as a good forum for further interaction and for the integration of local knowledge into modern ways of conserving and protecting the environment. Clarke-Ekong's study on festivals in contemporary Ghana fits with the findings of the current research (Clarke-Ekong, 1997).

A new position has recently been created in the chieftaincy institution–that of the *Nkɔsoɔhene. Nkɔsoɔ* literally means development and *hene* means a chief. *Nkɔsoɔhene* thus means a chief in charge of development. This innovation has ecological relevance, as one of the roles of the *Nkɔsoɔhene* is to ensure that the development that takes place does not neglect or undermine the sustainable use of natural resources in the area.

Yɛrepra Yadeɛ Ritual

A clear example of the environmental consciousness of Berekum people is provided by what they refer to as the *Yɛrepra yadeɛ* (literally 'we are sweeping away diseases') ritual. This ritual is performed to prevent an outbreak of a disease in the community. *Maame* Mercy (Personal Communication, 2 January 2012, Biadan) explains that it is performed when the leaders of a community have good reason to suspect that an epidemic is imminent within the community. When the people are certain about the imminent danger, the chief of the area tasks his *Ɔbaapanyin/Ɔhemaa* (Queen Mother) to act by collecting an egg from each household of the village. A day is then set for the ritual, on which every household will bring the refuse for that day and assemble at a designated point. A libation prayer will then be made by the *Ɔbaapanyin/Ɔhemaa*, after which all the women will go around the entire village three times holding some *nsommɛ* (*Costus afer*) leaves, and the refuse will be brought out amidst shouting:

Yɛrepra yadeɛ oo!
Yɛrepra yadeɛ oo!
Yɛrepra yadeɛ oo!
Literally translated as:
We are sweeping away sickness
We are sweeping away sickness
We are sweeping away sickness

After the third round, the *nsommɛ* leaves and some of the refuse will be deposited at all the entrances to the village. This deposition is referred to as *mpanpim* (literally translated as 'barrier'). The *mpanpim* is symbolic, blocking the impending disease from entering the village

I have already noted that Berekum ecological knowledge is not as effective as it used to be, as a result of a multiplicity of factors. However, traditional Akan and Berekum people had complex indigenous mechanisms for managing biodiversity that were influenced by their religion and culture long before the Convention on Biological Diversity (CBD), which was the result of the 1992 Rio de Janeiro 'Earth Summit' (Attuquayefio and Fobil, 2005). It is against this backdrop that African indigenous religion has been referred to as 'profoundly ecological' (Schoffeleers 1978). This stands in stark contrast to Lynn White's concept of Christianity, which he views as 'the most anthropocentric religion the world has seen' (1967:1205). However, many authors dispute this portrayal of Christianity (see, e.g., Johnson, 2000; Harrison 1999; Joranson and Butigan, 1984). Johnson (2000), for instance, questions how Christianity can be predicated to be the root of modern environmental woes when regions of the world in which Christianity's influence is very minimal are also facing environmental crisis. Harrison (1999) contends that a misreading or misinterpretation of Genesis, particularly1:27-28 led Lynn White and his supporters to connect Christianity with the wanton exploitation of nature. It is also interesting to note that long before Lynn White's statement, Aldo Leopold made a similar remark:

> Conservation is getting nowhere because it is incompatible with our Abrahamic concept of land. We abuse land because we regard it as a commodity belonging to us. When we see land as a community to which we belong, we may begin to use it with love and respect. There is no other way for land to survive the impact of mechanised man, or for us to reap from it the aesthetic harvest [that] it is capable, under science, of contributing to culture. That land is a community is the basic concept of ecology, but that land is to be loved and respected is an extension of ethics. That land yields a cultural harvest is a fact long known, but latterly often forgotten (Leopold, 1949: viii).

This revealing statement from Leopold did not, however, attract the kind of criticisms that Lynn White's attracted. Moreover, Milton (1998) argues that:

> People [indigenous people] who behave in non-destructive way that enable them to live sustainably do not necessarily respect their environment. [For] their material requirements may be such that they simply do not need to stretch their environment's capacity to support them. [...] To say that a

community has ethics or an ideology of respect for their environment is different from saying that they act in environmentally benign ways (Milton, 1998: 87).

However, using the case study of traditional Berekum, I have argued that conservationist considerations can be found in the beliefs and practices of indigenous or traditional peoples. Those who view the ecological values in traditional peoples' practices as mere coincidence, with the motivations for such practices deriving from religious considerations rather than conservationist ends, may have correctly identified the realities in *some* but not *all* indigenous societies. The next chapter examines in greater detail the impact of colonisation and Christianity on the conservation of nature in contemporary Berekum society.

Summary

I have thus far demonstrated that the relevance of methods for acquiring and transmitting knowledge within a society cannot be over-emphasised. I have pointed to the correlation between a society's capacity to develop and sustain itself and how it acquires and transfers knowledge. The analysis of the data confirms that the Berekum people's religious worldview helps them to live in harmony with nature, as studies of the religious worldviews of other indigenous societies have also found. However, there is also evidence that indigenous methods for dealing with environmental problems have been impacted on as the result of cultural contact and modernity, which have weakened the effectiveness of indigenous means of handling environmental issues. The above problems notwithstanding, the Berekum people's ways of creating environmental awareness and combating environmental problems provide a potential resource that can be tapped into, and refined where necessary, in order to complement scientific approaches to dealing with contemporary environmental problems. This is needed as historical evidence has shown that even with all its tremendous inroads, science has yet to adequately address the environmental problems of the world. This makes the synergy of both the indigenous and modern ways of tackling environmental issues an important end to pursue.

CHAPTER SEVEN

INDIGENOUS BEREKUM LIFE WAYS ENCOUNTER WITH COLONIALISM AND CHRISTIANITY AND ITS IMPACTS ON ENVIRONMENTAL MANAGEMENT IN BEREKUM

Introduction

This chapter focuses on the impact of colonisation and Christianisation on local environmental management in the study area. In order to discuss this, I will first need to look at how the experiences of colonisation and missionary activities have impacted on the institution of chieftaincy and traditional religion in Berekum. This is because the Chiefs are the supervisors of local environmental management, and traditional ecological practices are also underpinned by the indigenous religion. Therefore, anything that affects these two institutions is bound to affect environmental management in the area as well. There is evidence that both colonisation and Christianisation contributed to the decline of both the powers of Berekum Chiefs and the influence of the indigenous religion, which has implications for environmental management in the area.

A detailed account of the political life (chieftaincy) of the Berekum people was provided in Chapter Two, so here the focus is on how the encounters with colonialism and Christianity impacted on chieftaincy. As I have explained, a Chief is the living representative of the founding ancestors of the community in which he or she rules, and ancestors' consent is always sought through the medium of divination before a Chief is appointed. Thus, the position of a chief is sacred, and among the Berekum and the Akan in general, a Chief loses his or her legitimacy to rule if he reneges on his religious duties. This was confirmed by all the Chiefs that I interviewed, and it suggests that it is implausible to separate the influence of religion from the state in traditional Akan society, as political leadership is intrinsically fused with spiritual leadership. The legal basis of the authority for the traditional ruler is embedded in the religious mentality of the Berekum people (see Awuah-Nyamekye, 2009c).

The chiefs I interviewed stressed that there were very few problems involved in controlling the activities of the people during the pre-colonial era, as anyone who dared to challenge the authority of the chief was deemed to be challenging the authority of the ancestors. Also, all the laws, customs, taboos and other codes of ethics enacted by the Chief and his or her elders in the traditional communities were deemed to have divine backing, as they were believed to have been sanctioned by the *abosom* (gods) and *nananom nsamanfoɔ* (ancestors). This belief strengthened the enforcement of indigenous laws, including those regarding the environment. This set-up remained in place until the people of Berekum came under the combined influence of colonialism and Christianity during the early 20th century. I must, however point out that it is misleading to say that the environmental problems began with Christianity's introduction to Berekum in the colonial times, as available evidence indicate that it was only much later after independence that Christianity began to have a substantial influence in the area. However, my interviewees viewed colonialism and Christianity to provide the *basis* for the change in the Berekum people's attitude towards the indigenous ways of handling environmental issues.

The impact of colonialism on the traditional system of governance

The indigenous political system began to lose its grip on Berekum people during colonial rule. This is because the legislations that were passed by the colonial administrators affected everybody, since the colonial power–Britain–had gained total control of the administration of the country. The governors introduced a series of laws called *Native Jurisdiction Ordinances*, which granted the indigenous rulers (chiefs) only limited powers to preside over cases (Buah, 1998; see Awuah-Nyamekye, 2009c). In 1927, Sir Gordon Guggisberg (1919-1927), the governor at that time, introduced the *Native Administrative Ordinance*, which created the Joint Provincial Council of Chiefs, but this did not help much. Sir Alan Burns introduced a new *Native Authority Ordinance* in 1944, which was reintroduced in the governor's Constitution of 1946, and enabled a Native Authority to be appointed by the government. However, under the law, the decisions of the Native Authority could be reviewed or even set aside by the governor (Buah, 1998). Thus, although the policy of indirect rule involved the chiefs in the administration, the real power rested with the colonial administrators, since the authority of the chiefs operated under the general direction and control of the colonial district commissioners (Buah, 1998). Berekum was raised to paramountcy status by the colonial government of Ghana in 1901. This is an indication that the chiefs were essentially under the control of the colonial administrators.

The events leading to the establishment of the Catholic Church in Berekum also show the extent of the colonial administrators' powers. The records of the Saint Theresa's Catholic Church in Berekum show that when the church elders approached Nana Kwadwo Banie–the *Ɔmanhene* of Berekum at that time–to request suitable land on which to build their church, he refused. As a result:

> Msgr. Hauger and Fr. Fischer therefore went to Sunyani [the Provincial Capital] to see the governor, the District Commissioner (D.C). Major Robertson went to Berekum and told the Omanhene to give the Catholic mission as much land as he had given to the Protestants. The Omanhene heeded and gave the land. Fr. Fischer also provided a bottle of whisky, one pound and five shillings to the Omanhene, officially closing the deal (Saint Theresa's Catholic Church in Berekum, Records, 2009).

The analysis of the above drama, especially the giving of the drink and the payment of one pound and five shillings to the Chief, shows that the colonial administrators seemed to take the view that although the land belongs to the Chief, he has no real authority over it. This appears to give credence to my interviewees' contention that the Ghanaian tradition which makes chiefs the custodians of their lands was given little respect during the colonial era. This also shows that the colonial administrators worked in close partnership with missionaries to control the colonised.

Many of the Chiefs I interviewed stressed that although the disregard for the institution of chieftaincy began during the colonial era, it continued after the country's independence in 1957. This is evidenced by Ghana's post-independence history. Nkrumah's Convention People's Party (CPP) government did not depart from the way that the colonial administrators treated Chiefs after independence, and arguably devalued them even more, passing a number of parliamentary acts to reduce their authority between 1958 and 1961. The Ashanti Stool Act, 1958 (Act 28) and the Stool Lands Control Act, 1960 (Act 79) prevented chiefs from the regulation, collection and use of stool revenue. In the words of Odotei (2010), 'the Nkrumah government not only weakened the political and judicial roles of the chiefs but also made them appendages to the central government by breaking their financial backbone'. This has adversely affected the weal of the country, particularly in terms of its culture.

A good number of my interlocutors, most notably the Chiefs, were not happy with the way that Ghanaian politicians hold on to the legacy bequeathed to Ghana by the colonialists, particularly with reference to economic policies. For instance, economic considerations are placed higher than environmental considerations in all the major policies in the country. For instance, logging is increasing at an alarming rate, and mineral deposits are being overexploited throughout the country. As at 1998, the government of Ghana had granted over 200 mining leases to companies. This means that about 30% of Ghana`s land surface is being mined, implying that the volume of mining related hazards particularly, wastes and its attendant environmental degradation is growing at an alarming rate.(Awuah-Nyamekye and Sarfo-Mensah, 2012:.162).

The Chiefs complained that the central government rarely consults them when taking decisions that affect them and their subjects either directly or indirectly. A common grievance amongst the Chiefs concerned the award of timber concessions to contractors. A prominent Chief in the study area said:

> As a Chief, you would be there only to see a group of people on your land cutting timber. When you ask them they would produce a permit from the Forestry Commission authorising them to cut the trees. This is an affront on us as custodians of the land. You will complain but nothing serious comes out of it. The government is only interested in the taxes it gets from the contractors. This explains the fast depletion of economic trees in this area today (Personal Communication, 17 December 2011).

The situation is not peculiar to Berekum area, with Boon *et al.* (2009) raising the same point in a study in another region of Ghana. Kendie and Guri make a related point:

> Traditional institutions have always existed in Ghanaian society and have sustained society despite several years of external control by governmental institutions. State institutions tend to work outside of the traditional institutions. […] The consequence has been the neglect of enriching ideas from local practice (Kendie and Guri, 2006: 333).

My findings showed that, in reality, the institution of chieftaincy is now under more threat than ever in Ghana. For instance, the law that established Ghana's de-centralisation policy in 1988 stipulates that the Chiefs of the area concerned will be consulted by the President before District, Municipal and Metropolitan Chief Executives are appointed in their area, but in reality, this rarely happens. The Chiefs told me that they are usually cut off from such decision-making, and that things are even worse for chiefs who are viewed as being sympathetic to the opposition party. They said that this attitude has trickled down to the District, Municipal and the Metropolitan levels. The decentralisation policy says that one-third of Assembly members should be appointed to the Districts, Municipal or Metropolitan Assemblies by the government, and there is a specific clause in the law that says that Chiefs of the area shall be represented in the Assembly. However, the chiefs were unanimous in saying that the ruling party members at the local level usually select the Chiefs that they feel are 'in their camp', or sympathetic to the ruling government's policies and causes. This shows that although provision for de-centralisation exists in theory, it is meaningless for the Chiefs in practice.

Another important factor that Nana Professor Agyewodin Adu-Gyamfi Ampem (Personal Communication, 25 November 2012) made with reference to the threat to the authority of the chiefs was that Ghana's Supreme Court made a ruling in 2008 that said it is no longer obligatory for a subject to present him or herself before a Chief whenever he or she is summoned to appear before the court of a Chief. Interestingly this ruling was brought about through the actions of an ex-Chief and legal practitioner, Nana Adjei Ampofo, who filed a writ at the Supreme Court in September 2008 against the Attorney-General and the President of the National House of Chiefs that challenged the constitutionality of the Section 63 of the Chieftaincy Act (Act 759). In his writ, he stated that sub-sections b, d and e of Section 63–which stipulate that 'a person who being a chief assumes a position that the person is not entitled to by custom; deliberately refuses to honour a call from a Chief to attend to an issue; and refuses to undertake communal labour announced by a chief without reasonable cause; commits an offence and is liable on summary conviction to a fine, or to a term of imprisonment, or to both'–contravened Articles 14 and 21 of Ghana's 1992 Constitution (which deal with the Fundamental Human Rights and Freedoms of individuals). The Supreme Court upheld his objections, and this ruling has, seriously undermined and weakened the authority of the chiefs over their subjects.

The Chiefs were also unanimous in holding that their authority has been taken away by the modern system of government. This is confirmed by studies elsewhere in Ghana. For instance, Odotei and Awedoba (2006) observed that 'traditional states now operate within the modern state and conform to the structures that the state uses to regulate its operation'(2006: 17). As has been noted, the state's manipulation of the institution of chieftaincy began during the colonial era, with the indirect system of rule adopted by the British making chiefs appendages to imperial power (Awuah-Nyamekye, 2009c: 8). And this weakening of the authority of chiefs–particularly the transfer of decision-making from traditional authorities to the modern state-control institutions and agencies (see Migdal, 1974; Little and Brokensha, 1987; Venema, 1995)–has also occurred in other parts of Africa. Byers, Cunliffe and Hudak (2001, p. 194), quoting Derman (1996) on the chieftaincy situation in Zimbabwe, state that 'the enactment of the Communal Lands Act of 1982, and repealing of the Tribal Trust Lands Act of 1979, shifted land allocation authority from traditional political leaders to District Councils'. They added that, 'Traditional leaders have no legal means to compel modern leaders to respect traditional land use rules. The Chiefs have no legal power to create and enforce rules on natural resource management in communal lands.'

However, during my interactions with the Chiefs, it came to light that they were partially responsible for their current plight, as some Chiefs have been involved in corruption. For instance, some Chiefs mentioned that they know colleagues that are involved in multiple sales of land to developers. Some, they said, even connive with timber contractors to steal trees from sacred groves and other restricted areas. In the words of a prominent Chief, 'Chiefs are now not faithful to their duties, for the chieftaincy institution has been turned into a business venture.' Another disturbing issue that came out of the discussion was that fact that, as of now, some king-makers[1] could collude with wealthy people that don't belong to royal families and get them appointed as Chiefs for a price. Moreover, they reported that the procedures for selecting chiefs aren't always followed, even with those who are from the royal clans. All this undermines the authority of a chief. Chieftaincy disputes prevent the Chiefs from being able to present a common front for dealing with the challenges before them.

Colonialism and Christianity's impact on Berekum Traditional Religion

The Chiefs' loss of powers has had an effect on traditional religion's hold over indigenous people as the chiefs both embody and are the custodians of the indigenous religion. I havealready ointed out that, a

[1] This term is used to refer to the group of people who select and install Chiefs in Ghana.

significant body of legislation was passed outlawing African religio-cultural practices and, some of the policies the colonial administrators adopted paved the way for missionaries to launch their onslaught against the host religion. Major targets of the missionary assault included ancestral veneration, divination, and rites of passage. Smith remarks that 'the attitude of the missionary towards Akan religion and social customs was destructive [...] the Christian congregation came into existence in conscious opposition to the ancestral ways of life and thought of the rest of the community' (1986: 86). Juhe-Beaulaton (2008) expresses similar concerns in relation to the Benin and Togo people of West Africa.

Ɔkɔmfoɔ Agyemang mentioned that many aspects of indigenous religion were banned during that era, with laws being introduced against the practice of human sacrifice, the use of anti-witchcraft shrines, trial by ordeal, and *funusoa* (carrying the corpse to find out who was responsible for the death of a person) (Personal Communication, 11 November 2011). Debrunner corroborated this: 'Traditional justice could not be carried out as before. The custom of carrying the corpse and poison ordeal with *odom* bark was prohibited officially and could only be continued secretly' (Debrunner, 1961:. 63). In respect of the ban, Rattray (1959) reported that:

> In this modern example, typical of hundreds of such cases that once decimated whole villages, the tradition of centuries was so firmly instilled in the mind of the accused man that he seemed to have forgotten that he had only to appeal to the nearest European court to find redress (Rattray, 1959:167).

My informants acknowledged that not all the colonial bans were negative. For instance, they applauded the ban on human sacrifice, which they said was not acceptable for any reason. And banning trial by ordeal is certainly a step in the right direction as well.

The following from the report on *The Native Affairs Department*, January 1922 to March 1923, describes the ban on anti-witchcraft shrines:

> Owing to various reports from all parts of the colony and Ashanti clearly showing the distinctly harmful tendencies of the Fetish known variously as Hwenisu, Hwe Me So, Buani, Eguase, Kwana, Kwaku Mframa, Kwasa, Donkor, Brekum Akua, Sakra, the new cult, which had its stronghold in Akyem Kotoku, was suppressed by an Order – in – Council dated 9th December 1922 under the provisions of Section 14 of the Native Customs Ordinance. In January the Omanhene of Akyem Kotoku sent special messengers to Accra to protest against the suppression of this Fetish that he attributed to misrepresentations against him by his Political enemies acted

> on without his being afforded an opportunity of defending his division and their fetish (cited in Assimeng, 1985: 174).

Early literature on the missionary venture in Africa shows that missionaries came to Africa with a preconceived notion of the African. For instance, early European accounts of the Akan by authors such as Bowdich, Dupuis and Bosman created the impression that they were full of cruelty as a result of the practice of human sacrifice, and that they were also very fetishistic (cited in Onyina 2002: 134). Writing about the Presbyterian in Ghana, Noel Smith pointed out that the Basel missionaries were constantly promoting the view that the government should legislate against what they termed as 'heathenism' by prohibiting traditional practices. 'Heathenism', according to the Basel Mission, includes fetish worship, polygamy, the power of the chiefs, and funeral practices (Smith:. 89, cited in Onyina, 2002: 185). As I have already noted, the institution of chieftaincy is the embodiment of the Akan culture, and therefore attacking it implied that 'virtually everything of the Akan was evil and must be prohibited' (Onyina, 2002: 135).

The ban placed on anti-witchcraft shrines as well as some other traditional religious practices by the colonial powers not only affected the religious life of the traditional Ghanaian, but also the system of social control in traditional society. This is because traditionalists used divination as a mechanism for social control by utilising the widely held belief that diviners were capable of exposing evildoers and social deviants. Western ideas of social control were different from traditional African ones, and continue in some ways to influence post-colonial thought. This brief historical overview shows that there was a paradigm shift in Ghanaian's social practices and organisation as a result of their encounter with Western Europeans.

According to Nana Akumfi Ameyaw (Personal Communication, 7 November 2011), by attacking Ghanaian beliefs and practices, the missionaries were striking at the core of African culture, as 'African culture is a religious culture' and thus anything that affects its religion affects its culture in general. He said that Christians refused to comply with Chiefs on several grounds, and Parrinder (1961) produced similar findings in his study African societies. The effects of Christianity on African Traditional Religion and Culture are expressed concisely by Nukunya (1986):

> In order to understand the effects of Christianity on African culture, it is necessary to realise that traditional institutions never existed or functioned separately like discrete entities but dovetailed into one another. Religion

> for instance never operated independently of the kinship structure and the social control and vice versa. Any impingement on the religious beliefs, practices and organisations was, therefore, bound to affect not only that aspect of social life but also other elements of it (Nukunya, 1986: 87).

The impacts of impinging faiths on Berekum's traditional religion have been widespread. One traditional priest reported that he had not been able to celebrate his god's annual festival for years because his children, who used to help him during worship, have all become Christians. Some cults have ceased to exist and their shrines are now in ruins. Yaw *Kɔmfoɔ* (personal communication, 13 November 2011) remarked that formerly powerful shrines such as *Tano Kwasi, Tano Yaw, Tano Paamu, Brogya Kukuo, Diamono, Apomasu, Tigare, Nangro and Mframa*, which held centre stage in traditional African worship years ago, are now completely wiped out, either because there are not enough priests or not enough worshippers to assist the priests. This was confirmed at Kato by Nana Gyabaa, the *Abrafoɔhene* of Kato. He took me to places where two powerful shrines--*Ntoanfiri* and *Nana Gyabi*--used to stand and said:

> These two shrines had played a significant role in the development of Kato, but since their priests died quite some time ago nobody has shown an interest in becoming their priests because everybody claims to be a Christian. The shrines have thus been neglected for years and now they are in total ruins (Personal Communication, 11 December 2011).

He added that the only people who still demonstrate any zeal in the indigenous religion are elders.

Figure 20 Ntoanfiri shrine in ruins

Figure 21 Gyabi shrine in ruins

Most of the younger generation show little to no interest in the indigenous religion, and few will attend formal traditional worship. He claimed that the real challenge to Berekum traditional religion will thus come when its aged adherents pass away. As the table 6 in Chapter Four shows, there is only active cultic worship or attention at thirteen of the thirty-seven shrines listed there, with seventeen being dormant, and seven being completely unused. Thus, traditional religion–which forms the basis of traditional ecological practices–has been shaken to its very foundation. This means that the sacred groves which hitherto could not be entered by ordinary people, for instance, are now entered by many, and sometimes even logged by timber contractors. Because of the strong hold that Western culture and Christianity have on the people of Berekum, the use of the gods as police to enforce compliance and the ability of traditional leaders to guide young Berekums has been weakened.

One major factor which sustained membership in Berekum traditional religion was its priests' abilities to predict or reveal hidden things to its adherents. But today, many Christian prophets, prophetesses, evangelists, and Muslim clerics known as *mallams* (diviners) in the study area are seen as being able to do the same, leading many traditional Berekum religion believers to seek divine guidance from these priests and *mallams* instead. This has been another challenge to the sustenance of Berekum Traditional Religion in the 21st century.

One concern that almost all my key informants raised was that most of the young people in the study area that have received a Western education look down on traditional religion, with some even considering it as a religion of the illiterate folk. However, my informants noted that those with such opinions only say this when things are going well for them, and secretly solicit help from traditional priests when crises befall them.

The impact of Colonialism and Christianity on environmental management in Berekum Area

As I have noted earlier, the Chiefs in the traditional area are the main enforcers of indigenous conservation laws, and thus there is a correlation between the authority of a Chief and local environmental management. Many Chiefs and key informants I consulted maintain that the current environmental situation in Ghana can be directly traced back to the loss of the Chief's authority.

They contend that the colonial masters were to be blamed for the current over-exploitation of Ghana's natural resources, particularly its minerals and forest products. They argued that during the colonial era, the commercialisation of minerals (gold, diamond and bauxite) and timber became more pronounced in Ghana, and that these resources began to be exported in large quantities to the Western world. Cecile Jackson (1994) concurs: 'Early colonial policies were concerned with securing natural resources for European interests and introduced many environmental policies to achieve this end and regulate usage of natural resources' (cited in Amanor, 2001, p. 64). Obviously, the colonial masters' imposition of their way of life onto the African negatively affected indigenous people in many ways, in particular their ways of conserving nature. According to Koran-Amoako (Personal Communication, 11 October 2011), modern technology and industry create more environmental problems than they address. For instance, he cited the use of modern technology for logging, which has led to an increasingly rapid depletion of forest, and the use of chemicals and the creation of toxic by-products by industries, which are very harmful to both human beings and marine life. Similar opinions were expressed by almost all my respondents.

I have noted earlier that some of my informants acknowledged that colonialism has left some positive legacies in Ghana, such as the modern National Forest Reserves, including the Tain II and Pamu-Berekum Reserves in the study area. However, they were sad that the role the reserves were supposed to play had not been wholly realised because proper consultations were not made by the colonial administrators before the land was taken from the local people and this explains why the people enter these reserves for the least opportunity to cut trees or poach with the result that both are now severely depleted. This point to some weaknesses in the modern ways of conserving the environment, especially during the decision-making stage through to their implementation. Thus, while Berekums recognise some of the positive influences of colonialism, many of the older generations believe that the damage that has resulted from its impact on their culture is monumental.

My informants were unanimous in citing Christianity as a central cause of environmental degradation in the study area in response to the following question: *What do you think about traditional Berekum ways of conserving the environment and how effective are these ways of conserving nature today?* Almost every respondent said that the traditional methods had lost some of their effectiveness, and mentioned Christianity as a factor in this loss. The condemnation of Christianity was so overwhelming that it will be revealing to present some of the answers my informants provided for emphasis.

In a response to the above question, *Ɔkyeame* Okra Acheampong, the chief spokesperson of the *Akwamuhene* of Berekum Traditional Area said:

'Christians have no regard for traditions. They flout local environmental laws with impunity, particularly those of the Pentecostal type'. A typical example of this is the encounter between the Seventh Day Pentecostal Church and the *Ɔman* (state) I pointed out earlier on' (Personal Communication, 07 November 2011).

Figure 22. The researcher in an interview with Ɔkyeame Okra Acheampong

Yaw *Kɔmfoɔ* contended that Christian condemnation of the indigenous belief that spirits reside in trees, fields, mountains, streams and other natural objects as paganism or idolatry undermined an important belief that promoted the protection of nature (Personal Communication, 19 November 2012), and claimed that this has ecological repercussions. Lynn White also pointed this out: 'By destroying pagan animism, Christianity made it possible to exploit nature in a mood of indifference to the feelings of natural objects' (White, 1967:1205). Nana *Kɔmfoɔ* Serwaa Agyemang, a traditional African priestess added:

> The problem could be attributed to the spread of Christianity in this area. Muslims are not like the Christians. The Muslims generally respect the traditional laws. In those days there was respect for the elderly. Now there is no distinction between the elderly and the youth. Today, they will tell you "I attend Church or I am a Christian". That is, they are not under traditional authority and, for that matter, traditional religion (Personal Communication, 15 December 2011, at her shrine).

According to Kwabena Yeboah, 'Christianity condemns the old order but what it brought is also not followed' (personal communication, 10 October 2011), while *Ɔpanin* Yaw Wusu said: 'The youth of today do not respect our laws. They do what the land hates all in the name of Christianity' (Yaw Owusu, personal communication, 11 November 2011). Similarly, *Ɔpanyin* Kyere Kwame expressed his frustration thus:

> *Ɛneemafoɔ* [the youth of today] do not regard our tradition anymore. When *Kwasibroni* (the Whiteman) came with their Bible all our gods were rendered ineffective. They said our gods were false gods. People now challenge tradition with impunity. The youth do not respect the elderly.

> They see tradition as old fashioned (*Ɔpanyin* Kyere Kwame, personal communication, 6 November 2011).

Nana Okofo Dartey, the *Gyasehene* of Berekum Traditional Area--who is also a Christian--criticised some of the attitudes that Christians adopt in Berekum, arguing that:

> There are some Christians who go to the extent of condemning practitioners of African traditional religion. At times they go to the extent of insulting the chief priest. They describe anyone who practices African traditional religion as *kwasea* [a fool]. In my opinion this is not right and must be condemned (Nana Okofo Dartey, personal communication, 14 December 2011).

In an interview with the priest attached to the *Nyinakofi* sacred grove, he emphasised that:

> Christians do not respect we the *akɔmfoɔ* [traditional priests]. They do not even regard us as human beings. Ironically we are the very people they run to whenever they are in deep trouble. Christianity condemns traditional practices and in fact, Christianity has made traditional laws less effective today. In those days nobody could enter this forest without authority but now people fetch firewood from here (Kwame Agyemang, *Nyinakofi Kɔmfoɔ*, personal communication, 11 November 2011).

Bemoaning the current situation, a ninety year-old woman said:

> Many years ago --when a majority of us were attached to the religion of our forefathers--we did not have problems at all with the environment. There was plenty of rainfall for our crops and plenty of big trees and fresh air was in abundance. But now, the rainfall pattern has changed drastically. The rains do not come at the time that they are expected and when they come, they do not last as they used to. The worst of this is the excessive heat we are experiencing which has resulted in crop failure and strange diseases. This change in the weather has even affected our rivers. Those that used to be perennial are now seasonal. I blame this on the disrespectfulness that has characterised our society today. *Ɛneemafoɔ* are difficult to control. They have no respect for our customs and tradition. This started when *Kwasibroni* (the Whiteman) brought his school and religion to us. This has seriously affected the overall weal of our environment (*Maame* Ackoma, personal communication, 5 October 2011).

Nana Adwoa Takyiwaa, the queen mother of Senase–who is also a Christian–expressed her view that:

> Due to Christianity all the *tete* things [customs and traditions] are fading away. Now *bragorɔ* [puberty rites] are not performed. We Christians do not respect our traditional ways of life but traditional way of life ensures environmental responsibility. Muslims are doing better than us [Christians] in terms of respect for traditional values. They respect traditions more than we do (Nana Adwoa Takyiwaa, personal communication, 12 November 2011).

Throughout the interviews, my key informants expressed the view that Muslims in the traditional area generally respect the local laws better than Christian do, with a number of Christians even acknowledging this, as the above interview excerpts show. A Muslim scholar that I interviewed contended that 'the condemnation of traditional people's way of conserving nature was not justified [under Islam], for a careful analysis of these methods points out clearly that they are the way Allah wants us to behave towards the environment (Mujib, personal communication, 12 November 2011).

Madam Felicia Gyamfi, the spokesperson of Nana Siraa Ababio III (the Queen Mother of Kato) also lamented:

> I am a Christian. I belonged to the Pentecost Church but I am not happy at all with the way some Christians attack our customs and traditions, especially our indigenous religion. I believe this religion played a key role in ensuring environmental sanity some time ago, and it can still be of help today (Personal Communication, 9 December 2011, at Nana Siraa Ababio III's palace in Kato).

In a response to the above question, Mr. Augustine Gyedu, a forester in the study area, and also a Christian, said:

> Traditional methods, though associated with traditional beliefs, they are found to be sustainable. But traditional beliefs are gradually losing their values to Christian beliefs and are now less effective (Augustine Gyedu, personal communication, 12 November 2011).

The general view of my key respondents was that Christians, particularly those of the 'penteco-charismatic' strand, show disrespect towards traditional beliefs and practices, breaking various taboos such as *nkyida* (the observance of rest days as sacred days or Sabbath for the traditionalists) with impunity. At times, some of the activities of the overzealous Christians become inimical to social stability.

Two important views can be deduced from the above responses. The first is that the people of the study area feel strongly that Christianity is to blame for the current environmental woes of the area, as Lynn White (1967) argued to be the case. The second is that some of the Christians in the area are not comfortable with the way that other Christians there condemn the traditional beliefs and practices in the area.

This is important, as it can be seen that the thrust of the interviewees' criticisms of Christianity was not directed to Christianity *per se*–i.e. they did not hold Christianity up as an anti-environmentalist religion–but rather towards Berekum Christians for their attacks on traditional religion. Both Christian and non-Christian interviewees alike viewed these attacks to have negatively impacted on traditional ecological practices, as attacking traditional religion implies indirectly attacking traditional approaches to ecological management.

This apportioning of blame may not be as simple as my interviewees have presented it to be, however, and a more critical examination of these issues is needed. Certain questions still need posing, such as whether the local people can be absolved from blame given that it is they themselves that have adopted Christianity and are spreading it faster than ever before in the area. The available records, as was noted in Chapter Four, show that only two Christian missionaries–Rev. Parregaux and Rev. N. T. Clarke–visited Berekum, in September 1904, and these two alone paved the way for the establishment of the first Christian church (the Presbyterian Church) in the traditional area. And even here, Nana Kwabena Owusu–who was then *Ɔmanhene* (Paramount Chief) of Berekum–was reported to have aided in Christianity's introduction Berekum. The records show that he even sent one of his sons, Atta Kwadwo, and his nephew, Kwaku Agyemang to the school set up by the Presbyterian Church in Berekum (Centenary Celebration and Thanksgiving Service, 2006: 20). According to the history of the church in Berekum:

> … Rev. Parregaux, then based in Kumasi, had taken a similar trip to Berekum. His reception by Nana Kwabena Owusu, the *Omanhene* of Berekum Traditional Area, appeared friendly, so he felt Berekum was a fertile ground for the introduction of Christianity. On his return to Kumasi, and hearing the report from Nkoranza, he set off to Nkoranza, where he convinced Rev. N. T. Clark of the welcome awaiting him at Berekum (Centenary Celebration and Thanksgiving Service, 2006: 16).

After this, there is no record of Rev. Parregaux and Clark visiting Berekum again, so it can be assumed that it was the local people who accepted the faith and spread it in the area. The histories of the Roman Catholic and the Methodist Churches in Berekum also show that it was the indigenes of Berekum that introduced the churches there rather than expatriate missionaries. Another important finding is that the majority of the Chiefs I interviewed reported being Christians. Some said they even delegate certain rituals associated with their office that they view as contradicting their Christians beliefs. It may be questioned whether such delegation undermines a Chief's authority, since it had been pointed out that a Chief who refuses to perform the rituals associated with his office loses his/her moral right to remain in that office. There is also the problem of the tensions that exist between such chiefs and their kingmakers. If the adoption of Christianity is leading to the destruction of the environment, which is vital to the very existence of the people, then it is important to ask whether the people should continue to patronise and even expand their practise of it. Is this not a contradiction in terms?

Blaming religion for these problems would have been reasonable if Islam had been the culprit, as the local people have not shown any serious interest in Islam. After the demise of the first and the second generations of indigenous Berekum people who accepted Islam, their children generally repudiated it in Berekum. Although there is no literature to support this opinion, there is however, evidence from orature in the study area. This can be seen from the fact that there are no Muslims at all in some villages in Berekum, whilst a significant amount of people can still be found that bear Muslim names (indicating that their parents were Muslims). Almost all current practising Muslims in Berekum are non-natives, mostly from the Northern Ghana, where Islam has stronger roots. The Muslim communities are found in the *Zongo*[2] areas of the big towns in the region. There may be a number of reasons for this decline in indigenous Muslims, but these are outside the purview of this book. Although many Berekum people perceive Christians to be the major cause of the current environmental problems in the area, the evidence on the ground does not thus strongly support this conclusion, since Berekum people themselves have mentioned factors such as their over-reliance on scientific methods for conserving the environment, over-population, protracted chieftaincy disputes in the area, and a lack of patriotism.

Christianity and environmental conservation in Berekum

An important finding of the study is that the view of Christian pastors (especially those from the main line churches) towards the adherents of the traditional Berekum religion is changing. Those interviewed said that the

---

[2] Zongo is a term used to describe the communities which are predominantly inhabited by Muslims in Ghana.

traditionalists have kept their faith in indigenous religious beliefs and practices that are conservational in character, whereas Christians have not until recently begun paying serious attention to biblical verses that have conservational underpinnings. This suggests a delayed move towards adopting the view that as Calvin DeWitt (2000) contends, 'the Bible contains divinely inspired wisdom about creation, and if we pay attention to this text, we can learn to relate to other creatures as God intends' (quoted in Baumann *et al.*, 2011: 59). Rev. Kumi of the Presbyterian Church bemoaned the early missionary view that held everything about African religion was bad, including the description of the way that they interact with other elements of creation in terms of idolatry. Rather, he suggests, they should have investigated the philosophy behind the indigenous people's actions, and done away with this early missionary legacy long ago (Personal Communication, 6 October 2011).

So what contribution have Berekum Christians made towards environmental conservation there? When I discussed Methodist contributions towards the mounting environmental problems in Berekum with Rev. Acheampong, he said that although the general perception on the ground is that Christianity is to blame for Berekum's environmental woes, the fault that is attributed to it is exaggerated. He argued the church has not stood aloof, and has been doing what it can to eradicate the perception that the environment is unimportant, whilst also making some contributions to addressing environmental problems in the area. He mentioned that the Church now has what he calls a 'Social Gospel', in which the focus of the preaching is on environmental consciousness, and the priest preaches about the need to preserve and conserve the environment. According to him, the Church sees itself as having both social and religious responsibilities. This fits with DeWitt's view that:

> Authority over things belongs to the Author of those things: we have no authority to destroy what we ourselves did not create; destruction of a grand master's work by its onlooker, beholder, or curator may be a disgrace to their creator (DeWitt, 2000: 297).

Rev. Acheampong added that Methodist priests in Berekum also touch on indigenous ways of conserving nature during their sermons. He said that they de-emphasise the religious underpinnings of these indigenous ways, instead trying to rationalise the philosophies behind the indigenous methods. This, according to him, has gone down well with members of his congregation, and perhaps this represents a good starting point for integration.

Rev. Father Augustine Twum Obour of the Catholic Church in Berekum was also full of praise for the traditionalists' attitude towards the environment, and was not happy with the attitude of some Christians–especially those of the 'penteco-charismatic' churches–towards the indigenous religion. He stated that they usually advise the pastors of the charismatic groups not to directly attack the traditionalists. According to him, this advice is given during their Local Council of Churches (LCC) meetings. He mentioned that the Catholic Church now has numerous tree plantations in many parts of the country. Although he admitted that the plantations were set up for economic reasons, he claimed that they still help to restore order in the environment, since not all the trees from the plantations are harvested (Personal Communication, 11 January 2012). This suggests that some of the mainline churches' attitudes towards the practices of the traditionalists have changed–a change that may be due to their recognition that they have a crucial role to play in the conservation of nature, given the inroads the church has made into Berekum culture and lives. However, the attitudes of the Pentecostals remain somewhat problematic according to many of the interviewees, as they are very critical of the indigenous people's religious life, particularly with regard to their belief in spirits.

However, I learnt that some of the Pentecostalists have started rethinking their theologies with respect to their attitude towards the traditionalists. For instance, in my interaction with the Head Pastor of the Pentecost Church in Berekum, as well as with other junior pastors, it became clear that they are now not only seriously rethinking their theology, but they are also attempting to alter their members' attitudes towards the traditionalists, particularly with respect to their (traditionalists) ecological practices.

Another set of Christian practitioners that adopt a conservationist attitude in the study area is the Seventh Day Adventists (SDA). This is seen from their attitude towards certain game animals. The SDAs, like Muslims, strongly believe that it is a sin to eat a strangled animal, and also believe that it is wrong to eat a *tote boa* (an animal whose hoofs are not divided). Therefore, they do not eat or kill animals such as tortoises, flying squirrels, monkeys, hares and many others. For this reason, members of the SDA are generally not hunters, as traps cannot be used to discriminate between animals that are acceptable to catch and kill and those that are not. Moreover, SDA's will not eat all types of fish. Elder Ofori (Personal Communication, 11 December 2012) says that this is based on God's injunctions about animals that are to be eaten and those not to be eaten as found in Leviticus Chapter Eleven.

My findings also showed that despite over ten decades of Christianisation in the Berekum area, some of the traditional beliefs are still strong. For instance, the belief in the *tumi* (spiritual power) of plants like *homakyem* (*Spiropetalum heterophyllum*) and *odii* (*Okuobaka aubrevillei*) is still common. Also, as was noted earlier, very few Christians in Berekum would dare fish or eat a crab from the River Koraa because of their association of the *tumi* in these plants and rivers with demons (see Jackal, 2010)--a major focus of the 'penteco-charismatic' churches in the area. Also, only a few of my interviewees who were Christians said they would not mind killing or eating their totemic animal or destroying a totemic plant.

I have already pointed out that some Christians in Berekum are not happy with the way other Christians treat traditional religious beliefs and practices. Some of my key informants said they had no problem whatsoever with combining different faiths inasmuch as the marriage of faiths caters for both their spiritual and physical needs, just as Asamoah-Gyadu (2003: 3) has noted. This brings us to the issue of hybridity, which has become a common feature of the religious life of many people in the study area.

Hybridity

Hybridity is what some scholars refer to as 'syncretism'. I prefer 'hybridity' to 'syncretism' due to the negative connotations that syncretism has assumed today. For instance, some people see syncretism as involving confusion, but in the view of the people under study, there is no confusion involved in combining different faiths, with Islam and Christianity being seen as useful supplements to the indigenous religion by many Berekum people. My key informants reported having no problem with combining different faiths, inasmuch as the marriage of faiths caters for both their spiritual and physical needs. This is expressed in a local proverb that '*ɛnam dodoɔ nsei nkwan*' [plenty of meat does not spoil the soup]. Olupona aptly points out that 'African religious experience supports and encourages pluralism. Such eclectism produces an attitude of tolerance and peaceful co-habitation towards other traditions and cultures (Olupona, 2000, p. xviii). This mindset is embedded in the African worldview. Mbiti and Burleson have argued that Africans 'come out of African Religion but they don't take off their traditional religiosity. They come as they are. They come as people whose world view is shaped according to African Religion' (Mbiti and Burleson, 1986: 12 cited in Gathogo, 2007: 251). This is similar to Aylward Shorter's contention that the African Christian does away with 'remarkably little of his former non- Christian outlook.' (Shorter, 1975:.7 cited in Gathogo, 2007:.249). It is thus common to find a Christian or Muslim convert in the study area that fuses beliefs from these religions with his or her traditional religious beliefs. Some members of the community are even of the opinion that one may not be far from right in thinking that both Islam and Christianity can be perceived as an integral religious heritage of the people. For instance, it is very common to find items such as charcoal, pepper, and onions, in the rice, sugar and *gari* (a local staple food from cassava)–all these (charcoal, pepper, and onions) are believed by adherents of the Berekum religion to ward off evil spirits–being sold in the market by women who claim to profess religions other than indigenous Berekum religion. Explaining the basis of hybridity in the study area, an old man said:

> *Krakye*[3] *yɛn som ne yɛn amanmrɛ nyinaa yɛ adekrɔ. Nson nsonoye biara nnim. Saa nti na obi a wakɔyɛ Okristoni koraa no, ntumi ndwane mfiri abibisom ho no* (Personal Communication, 2011).

This is literally translated as:

> *Krakye* our religion and culture are seen as one thing. There is no difference between the two. This explains why it is very difficult for even an African Christian to run away from his indigenous religion.

The ecological importance of hybridity lies in the fact that, given that people have not wholly abandoned their traditional religious beliefs and values, and Christians are rethinking their theologies, it is still thus possible to ground ecological practices on traditional beliefs and attitudes in the study area.

Moreover, as it is implausible to claim that Christianity has completely destroyed the local Berekum belief system and its associated ecological practices (or, indeed, the local belief systems in other Ghanaian communities–see Berkes, 2008), there must be other factors that have and are playing a key role in the environmental problems of the area. And although Christianity is responsible for the disintegration of traditional African religious values, some Christian leaders (academic Christian pastors) are at the forefront of championing the recovery of a respect for traditional religious values. This is evidenced by the fact that books and articles on African indigenous religion and values have been authored by Christians. These are Christian pastors who work in this area within

[3] The local term for someone who has received Western education.

academic publishing, including Mbiti (Anglican, Kenyan); Idowu (Methodist, Nigerian); Gaba (Evangelical Presbyterian, Ghanaian); Osei (Methodist, Ghanaian); and Peter Kwasi Sarpong (Catholic, Ghanaian). No one can talk about the study of African Traditional Religion without recourse to works like *African Religions and Philosophy* (Mbiti); *Introduction to African Religion* (Mbiti); *The Concepts of God in Africa* (Mbiti); *African Traditional Religion: A Definition* (Idowu); *Olodumare, God in Yoruba Belief* (Idowu); *Ibo Sacrifice* (Cardinal Arinze); *West African Traditional Religion* (Asare); *Divine Revelation and Traditional Religion* (Bekye), *Scriptures of an African people* (Gaba). The contribution of nationalists that are also Christians, including Jomo Kenyatta and J. B. Danquah cannot be ignored either. For instance, on Akan indigenous religion, no one can ignore J. B. Danquah`s *Akan Doctrine of God.* The list goes on and on. What this hybridity and this refocusing of Christian attitudes relating to the importance of traditional values means is that there is the possibility, at least, of reviving some of the indigenous ways of protecting or conserving the environment that have indigenous religious underpinnings, and using these to complement scientific methods and approaches.

Summary

I have demonstrated from the foregoing discussion that the encounter between indigenous Berekum people, colonialism and Christianity has resulted in a change in the study area, particularly in the areas of politics, religion and environmental management. There is also evidence that the indigenous religion has shown resilience in some areas, and that this has manifested itself through the adoption of hybrid religious systems. Thus, the values of the traditional religion are not totally lost, for there can be a common ground for complementarism. That is, with the right approach, some traditional ecological practices with religious underpinnings can still be used to complement modern scientific approaches to conserving the environment. The indigenous Berekum peoples' ecological knowledge and practices can still contribute, however modestly, to fighting the current environmental problems that they face.

CHAPTER EIGHT

CONCLUSION

Summary and Conclusion

The central argument of this book has been that indigenous knowledge with religio-cultural underpinnings can help in addressing some of the contemporary ecological problems in Ghana. I have established that the natural environment in Berekum has been highly degraded, and several factors have been shown to be responsible for this. The people of the area are very much aware of their environmental problems, and would like to do something to address them. From the discussions, I summarise the salient findings of the book, as follows:

- That indigenous ecological knowledge is currently under threat;
- That indigenous knowledge provides a potential tool for addressing some of Berekum's current environmental problems.
- That indigenous ecological knowledge alone is inadequate to confront contemporary environmental challenges;
- That despite the importance of science and technology, they are not a panacea for dealing with the environmental problems of the world today.

I have also pointed out that many Berekum people are unhappy about the way in which indigenous knowledge is currently being handled. The threat to indigenous knowledge in Berekum has been attributed to several factors, but the key ones among them are the effects that the inroads of Christianity have had in the area, and westernisation by way of the increasing intrusion of central government into local affairs at the expense of local initiatives and autonomy. I will summarise the effects of these two factors in turn.

My informants viewed Christianity and its effects--Western education, science and technology--just as Lynn White argues, as being responsible for the decline of the positive impact of indigenous knowledge, particularly their environmental knowledge. However, this view as I have argued, is overly simplistic and idealistic (see Dikirr, 2008), since the evidence on the ground suggests it is exaggerated. For instance, factors such as over-population and its attendant effects and the intrusion of central government into local affairs have also played a significant role in the decline in the use of indigenous knowledge and practices. Moreover, the data have also shown that some of the current environmental projects being undertaken in the study area, such as tree planting, are faith-based initiatives, mostly being carried out by the Churches. Furthermore--and more importantly--the Churches are now rethinking their theologies with a view to encouraging their members to show more respect to traditional beliefs, especially those that are environmentally friendly. Many of my Christian informants denounced the way that some of their Christian colleagues criticise anything relating to traditional Berekum beliefs and practices, and feel that some aspects of traditional practices–especially traditional ecological practices–have something valuable to offer and thus need to be preserved. However, Christians who have some sympathy towards certain traditional beliefs and practices are now in a dilemma as to how to approach this need to preserve traditional values in the face of their Christian principles and the formidable competing values of modernity.

The government, on the other hand, is also blamed for seeing Western science as the sole remedy for the environmental conservation and management crises in Ghana to the exclusion of the use of indigenous conservation strategies, which should be seen as being complementary to those of science. Interviewees from Berekum seemed to echo Lynn White's view that 'more science and more technology are not going to get us out of the present ecological crisis until we find a new religion, or rethink our old one' (1967: 1206). However, negotiating these problems is not as simple as my informants think. One reason for this is the population rise in Berekum, which increased from 39, 374 in 1960 to 129,628 in 2010 (see Chapter Three). This puts the government in a dilemma, as it also has to attempt to find a suitable way to ensure food security and shelter needs, which require more land for development, and this has to be balanced against environmental conservation.

However, the fact that the government has such responsibilities does not entirely absolve it from all blame. There is evidence that right from independence in 1957, successive governments have been aware of the potential of indigenous knowledge in managing environmental issues in Ghana, but have failed to give indigenous knowledge serious consideration in their policies. For instance, there is no mention of indigenous knowledge in Ghana's 'National Biodiversity Strategy' (Ministry of Environment and Science, 2002). I have pointed out that the recurring view expressed during the interviews was that before Ghana's contact with Europe, the Ghanaians had a way of relating to nature in a sustainable way, but that their attitudes towards and ways of relating to nature and the use of natural resources steadily changed as colonialism gained root, with traditional ways of managing natural resources being ignored, demeaned and displaced by Western scientific models. It is important to add that my informants were not suggesting that nothing beneficial emerged from colonialism. They recognise that increased literacy, improved healthcare, shelter, and potable water, amongst other things, came about because of the colonial experience. But what they emphasise is that; 'the imposition of the colonial masters' way of life on the African negatively affected the indigenous peoples in many ways, and particularly with respect to the indigenous ways of conserving the environment.

One problem, however, is that our grasp of what constitutes indigenous knowledge is incomplete, since many of the traditional mechanisms are under threat or in a state of collapse, and have not even been studied exhaustively (Awuah-Nyamekye, 2009b). Attuquayefio and Fobil (2005) have argued that many environmental conservation initiatives have not been effective in developing countries–particularly in countries in Africa–due to the tendency to downplay the critical link between traditional and scientific approaches to conservation.

Though, scientific and indigenous ways of acquiring knowledge and transmitting it may be distinct, there is evidence that there is scientific underpinning in some of the indigenous ways of interacting with the environment. Menzies' study makes a similar finding as he strongly argues:

> '[I]t is important to recognize the strong underlying points of similarity between natural science and local knowledge [indigenous] systems in terms of the process of observation, inference, verification, and predication that is common to both modes of apprending the ecological systems within which human beings live(Menzies, 2006: 9)

Agrawal (1995a) has even argued that "finding clear demarcations between indigenous and Western knowledge is futile, given the failure of philosophers of science to find satisfactory verification criteria to distinguish science from nonscience" (cited in Berskes, 2008:.13). Again, Agrawal (1995b) further argues that "it is difficult to adhere to a view of indigenous and Western forms of knowledge being untouched by each other" (1995b)

These scholars are alluding to the fact that traditional environmental conservation strategies and those of science are not mutually exclusive so that one is preferred to the total neglect of the other. I have already argued that a careful analysis of traditional methods--reveal some scientific underpinnings to them. For instance, the use of resting days and ban periods among the Akan fishermen gives the sea or rivers some respite during which younger fish can grow. Although these rest or ban periods have religious undertones, they coincide with the time that the fish lay their eggs. Kroma (1995) blames the current ineffectiveness of indigenous beliefs concerning environmental conservation on Western education and values. The marginalisation may also be partly due to the view that indigenous knowledge is inferior, but this is contrary to the findings of many studies, which show that indigenous knowledge has the potential to be a resource for handling local environmental problems (Ntiamoa-Baidu, 2008; Sheridan and Nyamweru, 2008; Attuquayefio and Fobil, 2005).

Indeed, it is evident from the data that a religio-cultural practice like the preservation of sacred groves can be of some assistance in the conservation of the environment, particularly in the areas in which they are located. A report attributed to the Environmental Profile of Ghana (2006) claims that the numerous sacred groves throughout the country are mostly managed by local communities, and that the majority of them have no legal status. That is, the state does not support the local people by way of legislation to assist them in the enforcement of traditional ecological laws—religiously inspired environmental laws. This does not give the local communities a strong motivation to enforce local laws regarding the management of the groves.

My informants, who are the key stakeholders in relation to environmental issues in their area, have argued that neither indigenous nor the modern scientific methods are individually sufficient for addressing the current environmental problems in Ghana. This suggests that a new paradigm is needed to address the current ecological problems confronting Berekum people and Ghanaians in general. This new paradigm lies in a synergy of both the indigenous and scientific ecological approaches. This integration can work if the following among others are considered:

i) Policy-makers facilitating and becoming actively involved in a dialogue in which indigenous people, and particularly experts on indigenous ecological knowledge, are invited to contribute to the discussions for formulating policies on environmental conservation in the country.
ii) Government support for traditional authorities through legislation that helps them to enforce traditional ecological laws, such as the ban on entry into sacred groves, indiscriminate hunting, fishing and other restrictions.

iii) The development of proper collaboration between local authorities and the Environmental Protection Authority (EPA). This could be in the form of joint efforts by the traditional authorities and the District/Municipal/Metropolitan authorities to protect rivers, sacred groves and other natural resources.
iv) Closer collaboration between Agricultural Extension Officers and farmers on the benefits of so-called 'environmentally friendly' farming methods.
v) Non-Governmental Organisations (NGOs), both secular and faith-based, being encouraged to team up with local authorities in the quest to salvage nature.
vi) Policy-makers taking steps to include indigenous knowledge (particularly indigenous ecological knowledge) into the school curriculum at all levels of education so that indigenous knowledge can be understood and accepted in the country as a complementary resource for national development.

However, laudable as the above suggestions are, they would not be as easy to implement since numerous obstacles have to be addressed first. For instance, presenting a paper based on a UNESCO-sponsored project on Ghana to the International Workshop on the Importance of Sacred Natural Sites for Biodiversity Conservation held in Kunming and Xishuangbanna Biosphere Reserve in China, 17-20 February 2003, Thomas Schaaf stressed the important lessons that can be learnt from the conservation and management of sacred natural sites. Schaaf mentioned the following as important lessons and conclusions about the conservation and management of sacred natural sites revealed by the project:

- A sacred natural site is an important element linking nature and culture;
- A sacred natural site is often an anchor for cultural identity;
- A sacred natural site can constitute an effective means for environmental conservation as it is embedded in local and traditional belief systems;
- Sacred natural sites have great value for conservation ecology, for example:
  - as areas of high biological diversity,
  - as sanctuaries for rare or threatened species,
  - as sites that protect freshwater sources, and
  - as indicator sites showing potential natural vegetation in areas subject to environmental degradation (important for the

> restoration and rehabilitation of degraded ecosystems) (Schaaf, 2003: 15).

However, Schaaf (2003) also draws attention to the fact that cultural values are dynamic and can change substantially over time. Traditional belief systems can also change, resulting in the abandoning of sacred natural sites and the relinquishing of their associated protection by local communities. Nugteren (2009) puts the point succinctly:

> If the public interest in conservation is made too dependent on religious sentiments and mythical associations, instead of on common sense and a general awareness of responsibility for maintaining a precarious balance, then who could predict what would happen to ecological activism once the religiously inspired concern would dwindle for one reason or another? (Nugteren, 2009: 160).

Two important facts may be deduced from Schaaf's research. The first is that it confirms my informants' beliefs that their religio-cultural beliefs and practices have ecological importance. The second is that, as Nugteren (2009) has also shown, a people's indigenous religion can be used as a basis for their ecological practices. Therefore, as I have discussed in Chapter Seven, when Western culture and religion (Christianity in particular) gained roots in the study area, the basis for the host culture's ways of managing the environment began to lose its ground, and this led to a change in their perceptions, with many of the younger generations not being interested in traditional beliefs and practices. Moreover, similar perceptions are held by most Ghanaian policy-makers, although there is a growing belief worldwide that inputs from indigenous peoples are valuable for addressing matters such as environmental degradation and deforestation.

Thus, we need to reconsider the top-down approaches to both ecological and socio-economic issues by giving much importance to the bottom-up approaches. Such a move may succeed, as the local peoples will feel that they are partners in finding solutions to common problems and this may encourage them to open up. The value of this approach has already been recognised and evidenced in the activities of some international conservation organisations (see Tomalin, 2013: 178). For instance, as I have already pointed out (see Chapter Five), UNESCO has an initiative that seeks to find ways to integrate both local and scientific means of addressing environmental problems (particularly in the area of sacred groves and sites). The importance of these initiatives lies in the fact that the government can build upon them, as they have with the *Jaagbo* and *Anweam* sacred groves sponsored by UNESCO in the northern and eastern regions of Ghana.

It is notable that the literature on UNESCO projects also points to the challenges involved in such collaboration (see Barre *et al.*, 2008; Corbin, 2008; Ntiamoa-Baidu *et al.*:2003), which exist in the areas where conservation regulations are not undertaken in consultation with local people. In such cases, the local people tend to show little interest in collaborating, or even resist such projects as they feel that outsiders are trying to impose their will on something that has religio-cultural importance for them (the local people). The importance of the UNESCO initiatives is that, the challenges that have been identified from the implementation of the initiatives can serve as a guide for policy-makers to discover what they should avoid in order not to impede the progress of otherwise well-intentioned projects. Collaborating with locals over such issues can enable policy-makers to gain the confidence of the local people for smooth cooperation and integration. Such cooperation may be undertaken as:

> Sacred forests are the bedrock for people and communities' religious and spiritual beliefs and cultural identity. Hence, it is morally unjustifiable to remove the rights of peoples and communities to protect and manage such sites (Ntiamoa-Baidu *et al.*, 2003).

An encouraging sign that such cooperation or integration could be achieved in Berekum is the collaboration that currently exists between the people of Kato and the Environmental Protection Agency's (EPA) Brong Ahafo branch, where the latter has supplied over two million plant seedlings that have been planted around the *Mfensi* sacred grove to serve as a buffer zone to protect it. What still requires adding is the legal backing for this laudable venture. Also, a mention should be made of the NGO, the Ghana Association for the Conservation of Nature (GACON) of the Brong Ahafo Region, which has successfully partnered the community of Buoyem in conserving and managing the sacred groves of that area (Corbin, 2008).

One other obstacle is the institution of chieftaincy itself--he fulcrum on which indigenous environmental management revolves. This institution is currently riddled with disputes as wealthy people that are not from the royal clans can 'buy' a chieftaincy position by bribing king-makers. There are also some instances in which the procedures that have been laid down for selecting and installing chiefs are not properly followed (as seems to be the case in the study area), and this leads to unnecessary chieftaincy disputes. The attitude and behaviours of some of the Chiefs are also sources of worry, and this means that the very people who are supposed to be the rallying poin--and, for that matter, the main supervisors of indigenous knowledge--are found to be undermining their own authority by inappropriate behaviour, as I have pointed out in Chapter Seven. All these factors undermine the very institution that is supposed to guide and exemplify traditional authority, and also be the agent for development and resource management in the respective communities.

Moreover, there is an issue concerning the feasibility of my respondents' suggestions in the face of the highly globalised and complex world of today, which involves the market economy, over-population, science and technology. In addition,--and as I have noted earlier--Ghanaian political elites now heavily rely on Western scientific models for addressing environmental problems. Additionally, there is evidence that as far back as the early 1960s, an attempt by some African leaders to reinstate traditional values and practices through the movement of *Negritude* failed. However, I have argued that the philosophy of '*Negritude*' did not achieve its goals because either the concept was not clear to the masses or it was just premised on a protest built on the resentment of foreign rule. However, the proposed changes to Ghana's ecological values and approaches do not constitute a protest, and indeed border on being grounded in survival–for they offer protection against the destruction of the life-support system of the people. This may be the catalyst that can prompt stakeholders, particularly the government, into being pro-active on the issues now.

It may be contended that despite their limitations, the discussion so far and other studies somewhere support the idea that indigenous people still 'can provide ecological, economical, political and socio-cultural information necessary for conservation' (Stevenson, 2005; cited in Barre *et al.*, 2008: 31). Scholars such as Mukammuri (1995), Sheridan and Nyamweru; Mapara 2009, among others have contended that pre-colonial African were able to manage with considerable degree of success their environmental challenges before the imposition of the Western scientific strategies by the colonial governments which have been adopted hook line and sinker by post-independence African governments.

Using sacred groves and other sacred sites in Ghana as their basis, Ntiamoa-Baidu *et al.* (2003) for instance, have argued that in spite of the pressure on indigenous conservation practices, 'sacred forests and groves have survived for many years and continue to this day throughout the African region. [..] This shows the strength of spiritual values for the creation and maintenance of a particular sacred space or species and is a very powerful force for nature conservation.'

The above difficulties notwithstanding, I do not think it is an option to abandon the possibility of using indigenous knowledge to help to address current environmental problems, especially where scientific approaches to addressing said problems have been inadequate as a result of abuse and misapplication. That is, we should not throw the baby away with the bath water. Although I have pointed out that factors such as formal education, foreign religions and global capitalism pose a serious threat to the survival of indigenous ways of handling ecological problems, it is also true that despite the impact of Western culture and Christianity, the indigenous Berekum people have not entirely abandoned their traditional beliefs and practices, as the discussion has shown. This means the indigenous beliefs and practices still have an influence on Berekum people, and there is an increasing awareness there, as well as in Ghana as a whole (and even throughout Africa[1]) that local knowledge needs to inform policy decisions.

Moreover, one should not forget as we attempt to find solutions to the current problems, that every culture provides the means for its members to handle their pressing issues. The logical implication of this is that, as long as there are different cultures, there will be diverse approaches even to similar problem(s) just as Sutton and Anderson (2010) have pointed out.

It has been clear from the discussions so far that environmental policies in Ghana have failed to achieve their desire goals because; all the policy-decisions makers failed to factor in the perspectives of the local people. In the words of oneof my key informants:

> Any environmental conservation policy document that does not recognise the input of the local people is bound to fail, for the local people are those who are on the ground and not the initiators of the policies who sit in their offices in Accra [the national capital]' and direct affairs.

The above statement is supported by the fact that there is worldwide evidence that Western science alone has not been able address the mounting environmental problems across the world. This is an indication that there is need for complementing strategies such as those from the indigenous knowledge systems of various groups across the world.

The foregoing discussions also have implications for those of us in the field of religion and the environment or Environmental Humanities in general. That is, it places a responsibility on us to facilitate this joint venture by explaining the issues to the key stakeholders and at the same time lobbying policy-decisions makers on the need to integrate indigenous

[1] Dikirr (2005) although critiquing the romanticisation of indigenous ecological knowledge, also accepts its potentials.

knowledge into their policies and implementation in the fight against the current environmental problems of the country. The lobby should also include getting policy-decisions makers to accept the need to include indigenous knowledge, particularly indigenous ecological knowledge into the school curriculum at all levels of education as my informants have suggested. Dikirr has made a similar suggestion that:

> What scholars and students of Africa need to do is to figure out how a creative synthesis between Africa's village-level of cooperative ethic, system of local governance, indigenous technologies of managing natural resources and Western techno-scientific skills might be realized' (2008, p. 96)

In conclusion, the analysis of the research data suggests that although we should be cautious of the romanticisation of the potential for indigenous knowledge to address environmental problems, and consider the complex nature of attempting to integrate indigenous knowledge with scientific conservation methods, the evidence on the ground makes it difficult to ignore the potential contribution that indigenous people could make to environmental conservation agendas. Also, and more importantly, as it stands now, neither modern (scientific) nor traditional modes of addressing current ecological problems are adequate to do the task required from them, and it is against this backdrop that the synergy of the two modes of addressing environmental problems becomes imperative. A case in point is where government legislation alone could not stop people from using dangerous chemicals for fishing, but the Chiefs and elders were able to by pronouncing curses on potential offenders.

This means that stressing the similarities between the two systems for conserving environment (indigenous and modern scientific) rather than their differences may provide the best way to begin considering how best to integrate these two diverse systems of knowledge. What this means is that there is the need to design a strategy, supported by the needed funding to undertake further research with the view to gaining a better understanding of indigenous peoples' conservation practices. This should be directed at exploring which elements of indigenous ecological knowledge and practice are applicable today, and the best ways to remove the obstacles to their application in order to pave the way for the integration of the two epistemologies that are required to successfully address the contemporary ecological problems of Ghana, and perhaps of Africa as a whole.

BIBLIOGRAPHY

Abayie Boaten, A. "Traditional Conservation Practices: Ghana's Example." *Institute of African Studies Research Review,*14,(1998): 42-51.

"Abrahamic Concept of Land." Accessed 14 June 2013. http://memexplex.com/Meme/2282/

Acquah, Francis. "The Impact of African Traditional Religious Beliefs and Cultural Values on Christian-Muslim Relations in Ghana from 1920 through the present: A Case Study of Nkusukum-Ekumfi-Enyan Area of the Central Region". PhD thesis, University of Exeter, UK., 2011

Adarkwa-Dadzie, A. "The contribution of Ghanaian Traditional Beliefs to Biodiversity Conservation." In *The Third UNESCO MAB Regional Seminar on Biodiversity Conservation and Sustainable Development in Anglophone Africa (BRAAF), 9-12 March 1997, Cape Coast*. edited by D.S. Amlalo, L.D. Atsiatorme and C. Fiati. Accra: Environmental Protection Agency (EPA), 1998.

Adarkwa, K. "Spatial Analysis of Rural Development Policies in Upper Region of Ghana." *Rural Africana*. (1983): 5-18.

Adimado, A. A. and Amegbey N. A. 2003. Incidents of Cyanide Spillage in Ghana, Mineral Processing and Extractive Metallurgy. (Trans. IMMC) 112, 2.

Adjewodah, P. and Beier, P. 2004. "Working with Traditional Authorities to Conserve Nature in West Africa." African Conservation Telegraph (DRAFT). Accessed 29 June 2012. http://www2.for.nau.edu/research/pb1/Service/AdjewodahBeier_Traditional_Authority_Conservation.pdf.

Adogame, Afe, Chitando, E and Bateye, E. *African Traditions in the Study of Religion in Africa*. England, USA: Ashgate. 2012.

Agar, M. H. and. Hobbs, J.R. "How to Grow Schemata Out of Interviews." In *Directions in Cognitive Anthropology* edited by J. W. D. Dougherty, 413-431. Urbana: University of Illinois Press, 1985

Agar, M H. "The Right Brain Strikes Back." In *Using Computers in Qualitative Research* edited by , N. Fielding and R. Lee, 181-194. Newbury Park: Sage Publications, 1991.

Agboro, O.P. 2008. "African environmental ethics: A creation of distorted values" (A paper presented at Rebuild Africa conference, Washington D.C, 8-9 August 2008. Accessed 12 January 2012.

http://www.scribd.com/doc/12529433/African-Environmental-Ethic-A-Creation-of-Distorted-Values-by-OMAM-PAT-AGBORO.

Agyarko, Tabi. "Forestry Outlook Study for Africa (FOSA) Country Report – Ghana." Accessed 20 April, 2012. http://www.fao.org/docrep/003/ab567e/AB567E00.htm, 2001.

Agyemang, Yaw.Sarkodie. "The Influence of some Aspects of Akan Religious Thought on the Lives of the Akan: A Case Study of the Sekyere" M.Phil. Thesis, University of Ghana, 1994.

Akabzaa, T.M., Banoeng-Yakubo, B.K. and Seyire, J.S. "Heavy Metals in Some Mining Communities Within The Jimi River Basin in Ashanti Region, Ghana." *Journal of Ghana Science Association*, 7 (2005): 36-45.

Akabzaa, T. M. and Darimani, A. Impact of Mining Sector Investment in Ghana: A Study of the Tarkwa Mining Region. A Draft Report prepared for Structural Adjustment Participatory Review International Network (SAPRIN), 2001.

Akinsanmi, O. *Certificate Agricultural Science*. London: Longman, 1975.

Akowuah, D. K, Rice K, Merz A and Sackey V.A "The Children Of The Gods." *Journal of Ghana Wildlife Society*,1 (2) (1975): 19–22.

Akyeampong, Emmanuel. Indigenous knowledge and maritime fishing in West Africa: The case of Ghana. *Tribes and Tribals, Special Volume* 1(2007):173-182.

Alley, D. Kelly. "Idioms of Degeneracy: Assessing Ganga's Purity and Pollution." In *Purifying the earthly body of God:Religion and Ecology in Hindu India*, edited by L. E. Nelson, 297-330. 1998.

—. "Separate Domains: Hinduism, Politics and Environmental Pollution." In *Hinduism and Ecology,* edited by C. Chapple and M.E.Tucker, 355-387. Cambridge MA: Harvard University Press, 2000.

—. *On the Banks of The Gange: When Wastewater Meets Sacred River*. Ann Arbor: University of Michigan Press, 2002.

Altheide, D.L. "Ethnographic Content Analysis." *Qualitative Sociology*, 10 (1987): 65-77.

Alolo, N A. "African Traditional Religion and the Concept of Development: A Background Paper.", in Birmingham: Religions and Development Research Programme Working Paper no 17.62 , 2007.

Al Qaradawi, Y. The Lawful and The Prohibited In Islam. 10900 W. Washington: America trust Publications, (not dated).

Amanor, Kojo Sebastian. *Land, Labour and the family in Southern Ghana: A Critique of Land Policy under Neo-Liberalisation*. Goteborg, Sweden: Elanders AB., 2001.

Anane, Mike. "Implementing Agenda 21: NGO Experiences From Around The World." In *United Nations Non-Governmental Liaison Service* (NGLS) edited by L. Alyanak, and A. Cruz, 99-107,1997.

Anane, Mike. Implementing Agenda 21: Religion and conservation in Ghana.Accessed 31 May 31, 2012 http://www.un-ngls.org/docments/publications.en/agenda21/12.

Anthwal, A., Sharma, R. C. and Sharma, A. "Sacred Groves: Traditional way of Conserving Plant Diversity" In Garhwal Himalaya, Uttaranchal. *The Journal of American Science*, 2(2)(2006): 35-38.

Appiah-Opoku, Seth. *The Need for Indigenous Knowledge in Environmental Impact Assessment: The case of Ghana.* New York: Edwin Mellen Press, 2005.

Appiah-Opoku, Seth and Hyma, B. "Indigenous Institutions and Resource Management in Ghana." *Indigenous Knowledge and Development Monitor*, 7(3) (1999):15-17.

Appiah-Opoku. Seth and Mulamoottil, G. "Indigenous Institution and Environmental Assessment: The case of Ghana." *Environmental Management, 21(2)* (1997):.159-171.

Arhin, S. Complementing legislation: The Role of Cultural Practices in the Conservation of Wildlife – Examples from Ghana. In Journal of Animal Law, 11(v) (2008): 93-98.

Arhin, K. "Asante Security Posts in the North-West: Seminar on Asante and Brong-Ahafo." Legon, Accra: Institute of African Studies, University of Ghana, 1973.

Asante, C. "Folktale And Environmental Education." In *Cahiers du Centre d'Etudes et de Recherche en Lettres Sciences Humaines et Sociales* (CERLESHS) edited by A. Napon , 101-112,5e numero special, 2005.

Asante-Darko, K. "Ecocriticism and Environmental Knowledge Of Asante Oral Traditional Poetry" PhD. Thesis. University of the Witwatersrand, Johannesburg, South Africa, 2006.

Asante, M.K. "Afrocentricity and Culture." In *African Culture: The rhythms of Unity*. edited by M.K Asante, and K. W Asante , 1-12. Trento, NJ. and Asmara, Eritrea: African World Press, Inc, 1996.

Asamoah-Gyadu, Kwabena. J. "Blowing the Cover: Media Projections of Religious Functionaries/Specialists in an African context." *Legon Journal of the Humanities, Faculty of Arts* , University of Ghana, 14(2003): 1-20.

Asiama, O. S. "Comparative Study Of Land Administration Systems: Case Study—Ghana." Accessed 2 April 2, 2012. Available http://www.landcoalition.org/sites/default/files/legacy/legacypdf/wbtghana.pdf.,1-40, 2007.

Assimeng, M. *Religion and Social Change in West Africa: An Introduction to Sociology of Religion.* Accra: Ghana Universities Press, 1985

Attfield, R. *Environmental Ethics: An Overview Of The Twenty-First Century.* Cambridge UK: Polity Press, 2010.

Attuquafio, Daniel.K. and Fobil, Julius .N. "An Overview Of Biodiversity Conservation In Ghana: Challenges and Prospects." *West African Journal of Applied Ecology, 7*, (2005):.1-18.

Attuquayefio, Daniel. K and Gyampoh, S. 2010, The Boabeng-Fiema Monkey Sanctuary, Ghana: A case for blending traditional and introduced wildlife conservation systems', *West African Journal of Applied Ecology*, 17, pp.1-10.

Alvesson, M and SkoldberG. K *Reflexive Methodology: New Vistas for Qualitative Research.* London: Sage, 2000.

Awolalu, J. O. "Sin and Its Removal in African Traditional Religion." *Journal of American Academy of Religion* XLIV (2) (1976): 275-287.

—. African View of Man., *Orita: Ibadan Journal of Religious Studies* 1(2)(1972):.101-118.

Awuah-Nyamekye, Samuel. "Belief In *Sasa*: Its Implications for Flora and Fauna Conservation in Ghana." *Nature and Culture*.7 (1) (2012a):1-15.

—. *Divination: Its Nature and Meaning in Berekum Traditional Area.* Verlag: Lap Lambert Academic Publishing Gmbh and Co, 2012b

Awuah-Nyamekye, Samuel, Totemism, 'Akyeneboa' and Plant Ethics. *PAN: Philosophy Activism Nature*, 9 (2012c): 5-10.

Awuah-Nyamekye, Samuel, and Sarfo-Mensah, P. "Mining Or Our Heritage? Indigenous Local People's Views On Industrial Waste Of Mines In Ghana." In *Industrial Waste* edited by S. Kuan-Yeow and G. Xinxin, 151-172. Rijeka, Croatia: In Tech., 2012.

Awuah-Nyamekye, Samuel, and Appiah-Sekyere, Paul.2012. "Teenage Pregnancy in the life and Thought of the Akan: Moral Perspectives." *Sociology Study*. 2 (2) (2012):129-138.

Awuah-Nyamekye, Samuel, and Sarfo-Mensah, Paul. 2011. Ensuring Equitable Distribution Of Land In Ghana: Spirituality Or Policy? A Case Study From The Forest-Savannah Agroecological Zone Of Ghana. *The International Indigenous Policy Journal*. [online]. 2 (4) (2011):1-20. Accessed October 21, 2011. Available http://ir.lib.uwo.ca/iipj/vol2/iss4/9.

Awuah-Nyamekye, Samuel. Salvaging Nature: Akan Religio-Cultural Perspective. *Worldview: Global Religions, Culture and Ecology*. 13 (3) (2009a): 251-282.

—. "Teaching Sustainable Development From The Perspective Of Indigenous Spiritualities Of Ghana." In, *Religion and Sustainable Development Opportunities and Challenges for Higher Education* edited by C. De Pater and I. Dankelman, 25-39. Berlin: Lit Verlag, 2009b.

—. "The Role of Religion in the Institution of Chieftaincy: the Case of Ghana." *Lumina*. 20 (2) (2009c):.1-17.

Bamfo, R. K. "Rights, Forests and Climate Change." *Climate and Forest Conference,15-17 October, Oslo, Norway.2008*

Banwari. Pancavati: *Indian Approach to Environment* (Tr. Asha Vohra), New Delhi: Shri Vinayaka Publications, 1992.

Barnett, E and Casper, M. "A Definition of Social Environment" *American Journal of Public Health*, 91, (3) (2001): 465.

Bartle, P. F. W. Forty Days: The Akan Calendar. *Africa. Journal of the International African Institute*.48 (1) (1978): .80-84.

Barre, R. Y., Miriam, G. and Dianne, D. 2008. "The Role of Taboos in Conservation of Sacred Groves in Ghana's Tallensi-Nabdam District.". *Social and Cultural Geography* 10 (1) (2008):.25-39.

Battiste, M. 2005. Indigenous Knowledge: Foundation for First Nations. Accessed May 7, 2012. http://www.viu.ca/integratedplanning/documents/IndegenousKnowledge PaperbyMarieBattistecopy.pdf, 2005.

Bascom, A. William. "The forms of folklore: Prose Narrative." *Journal of American Folklore,* 78(307) (1965): 3-20.

Bauman, Whitney A, Bohannon II, Richard R and O`Brien, Kevin J. (eds.). *Grounding religion: A field guide to the study of religion and ecology*. London & New York: Routledge, 2011.

Beavis, M. A. "Environmental Stewardship: History, Theory and Practice." Winnipeg: Workshop Proceedings, Institute of Urban Studies, University of Winnipeg, 1994.

Benneh, Kwaku. *History of Berekum: Political, Eeconomic and Social Aspects*.Undergraduate Thesis, University of Ghana, Ghana, 1990.

Berekum Traditional Council Records. 1949. History of Berekum.

Berekum District Assembly Survey, 2002.

Berekum Municipal Health Report, 2007.

Berekum Municipal Assembly Report, 1996

Berkes, Fikret. *Sacred Ecology* 2nd ed. New York & London: Routledge, 2008.

Biggelaar, C. D and Gold, M. A. "The use and Value of Multiple Methods to Capture the Diversity of Endogenous Agroforestry Knowledge: An example from Rwanda." *Agroforestry Systems* 30 (3) (1995):263-275.

Billig, M. "Rhetorical and discursive analysis: How Families talk about Royal Family." In *Doing Qualitative Analysis in Psychology,* edited by N. Hayes, 39-54. Sussex: Psychology Press Ltd., Publishers, 1997.

Blaut, J. M. *The colonizer's Model of the World: Geographical Diffusionism and Eurocentric History*. New York: Guildford Press, 1993.

Blay, D, Dwomoh, F and Damnyag, L. Case Studies on Measuring and Assessing Forest Degradation: Assessment of Forest Degradation by Local Communities: The Case Study of Ghana. Forest Resources Assessment Working Paper, 160 (2003):.1-24.

Boafo-Arthur, K. ".Chieftaincy and Politics in Ghana since 1982." *West Africa Review*. 1 (3) (2001):.13-17.

Boakye, A. Philomena. Key Note Address World Environmental Day, 5 October 2011. Kato near Berekum, 2011.

Boersema, J. J. "Environmental Sciences, Sustainability, and Quality." In *Principles of Environmental Sciences*, edited by J. J. Boersema and L. Reijnders. London: Springer, 2009.

Booth, J.S. "An Approach to African Religion." In *African Religions, Asymposiu,* edited by J.S. Booth . New York: NOK Publishers, 1997.

Boon, E, Ahenkan, A and Nakuku-Baduon, B. "An assessment of Forest Fesources Policy and Management in Ghana. Impact Assessment and Human well-being". 29th Annual Conference of the International Association for Impact Assessment, 16-22 May 2009, Accra International Conference Centre, Accra, Ghana.

Brightman, R.A. "Conservation and Resource Depletion: The Case of the Boreal Forest Algonquians." In *The Question of the Commons: The Culture and Ecology of Communal Resources*, edited by B.J. Mccay and J.M. Acheson 121-141.Tuscan: The University of Arizona Press, 1987.

Britannica – Encyclopaedia. "Myth". Accessed 12 May 2012. http://www.britannica.com/EBchecked/topic/400920/myth., (undated).

Brockington, D and Schmidt-Soltau, K. "The Social and Environment Impacts of Wilderness and Development." *Oryx*. 38(2005):140-42.

Brodt, S.B. "A Systems Perspective on the Conservation and Erosion of Indigenous Agricultural Knowledge in Central India." *Human Ecology* 29 (2001): 99-120

Brown, K. "Medicinal Plants, Indigenous Medicine and Conservation of Biodiversity in Ghana." Centre for Social and Economic Research on the Global Environment University of East Anglia and University College London (CSERGE)" Working Paper GEC 92-36, 1992.

Brundtland Report: Our Common Future. [Accessed 24 June 24, 2012 from:

http://www.are.admin.ch/themen/nachhaltig/00266/00540/00542/ index.html?lang=en, 1987.

Bryant, A. "Re-grounding GroundedTtheory." *Journal of Information Technology Theory and Application (JITTA)*: 4(1) (2002), Article 7.Accessed January 12, 2012 http://aisel.aisnet.org/jitta/vol4/iss1/7.

Bryman, A. E. "Trangulation" Accessed 28 April 28, 2011 http://www.referenceworld.com/sage/socialscience/triangulation.pdf, 2004

—. *Social Research Methods 3rd ed*. Oxford: Oxford University Press, 2008.

Buah, F. K. *A History of Ghana Revised and Updated*. London: Macmillan Educational Ltd., 1998.

Burley, H and Haslam, C. *How can I stopCclimate Change: What is it, Why it Happens and What you can do to help. London*: HarperCollins Publishers, 2008.

Busia, K. A. 1968 *The Position of the Chief in the Modern Political System of Ashanti*. London: Frank Cass & Co. Ltd, 1968 2nd impression [1951].

Byers, Bruce. A., Cunliffe, Robert. N. and HUDAK, Andrew. T. "Linking the Conservation of Culture and Nature: A Case Study of Sacred Forests in Zimbabwe". *Human Ecology,* 29 (2)(2001):187-218.

Carboo, D. Serfor-Armah, Y. "Arsenic in Stream and Sediments in Obuasi area. proceeding of the symposium on the mining industry and the environment." (KNUST/IDRC) (1997): 114 - 119.

Carson, R. *Silent Spring.* Boston: Houghton Mifflin, 1962.

Charmaz, K. *The Grounded Theory Method: An Explication and Interpretation. In Contemporary Field research: A Collection of Readings,* edited by R. M. Emerson, 109-128. Boston: Little, Brown and Company1983.

Chittick, W.C (ed.) [Forward by Huston Smith].*The Essential Seyyed Hossein Nasr*. Bloomington, Indiana, World Wisdom Inc., 2003

Chouin, G.Sacred Groves in History: Pathways to the Social shaping of Forest Landscapes in Coastal Ghana. *IDS Bulleting* 33(1) (2002):39-46.

Clarke, E. "The Sociological Significance of Ancestor–Worship in Ashanti, Africa." *Journal of the International African Institute*, 3(4) (1930): 431-471.

Clarke-Ekong, S. F. "Traditional Festivals in the Political Economy: The Case of Contemporary Ghana." *Journal of Social Development in Africa*, 1(2) (1997): 49-60.

Colding, J. and Folke, C. 1997. "The Relations Among Threatened Species, their Protection, and Taboos." *Conservation Ecology,* 1.6., 1997. Accessed March 10, 2012.
http://www.consecol. org/vol1/iss1/art6/.

Conservation International- Ghana. *Handbook of Totems in Ghana: Traditional Mechanism for Biodiversity Conservation,* Accra: Innolink, 2003.

Corbin, A. "Sacred Groves of Ghana." Accessed August 9, 2013.
http://www.sacredland.org/sacred-groves-of-ghana/, 2008.

Cox, J L. *An introduction to the Phenomenology of Religion*. London, New York: Continuum, 2010.

Cox, P.A. "Will Tribal Knowledge Survive the Millennium?" *Science*, 287(5450) (2000):.43-45.

Cooper, David, E and Palmer, Joy A. (eds.). *The Environment in Question: Ethics and Global Issues*. London and New York: Routledge, 1992.

Cupples, J and Kindon, S. "Returning to University and Writing the Field." In *Development Fieldwork,* edited by R. Scheyvens and D. STorey, 223. London: Sage Publications, 2003.

Danfulani, U.H.D .Africa Religions in Scholarship: A critique. In 2012. *African Taditions in the Study of Religion in Africa*, edited by A, Adogame,.E, Chitando and B. Bayete, 19-34. England, USA: Ashgate, 2012.

Danquah, J.B. *The Akan Doctrine of God*. London: Frank Cass, 1968.

—. *Akan Laws and Customs and the Akyim Abuakwa Constitution.* Oxford: Oxford University Press, 1928.

Debrunner, H. *Witchcraft in Ghana: A Study on the Belief In Destructive Witches and its effects on the Akan Tribes*. Accra: Presbyterian Book Depot,1961.

De clerck, F., Ingram J and Rumbaitis D. R. C. . "The role of Ecological Theory and Practice in Poverty Alleviation and Environmental Conservation." *Frontiers in Ecology and the Environment.* 4 (2006): 533-540.

Dercher, J. "Conservation, Small Mammals, and the Future of Sacred Groves in West Africa." *Biodiversity and Conservation* 6 (1996): 1007-1026.

Derman, B. *Changing Land-use in the Eastern Zambezi Valley: Socio-Economic Considerations*. CASS-WWF Joint Paper. Centre for Applied Social Sciences, University of Zimbabwe, P.O. Box MP 167, Mount Pleasant, Harare, Zimbabwe, 1996

Descola, P. and. Palsson, G. (eds.). "Introduction", *Nature and Society: Anthropological Perspectives.* London and New York: Routledge, 1996.

DeWitt, B. Calvin. "Behemoth and Batrachians in the Eyes of God: Responsibility to Other Kinds in Biblical Perspective." In *Christianity and Ecology: Seeking the Well-being of Earth and Humans*, edited by D. Hessel and R. Ruether,291-316 Cambridge, Massachusetts: Harvard University Press, 2000.

Dickson, Kwasi, and Ellingworth, Paul (eds.). *Biblical Revelation and African Beliefs*. London: Lutter Worth., 1969.

Dikirr, P. M "The Challenges of Indigenising Africa's Environmental Conservation Goals." *The Journal of Pan African Studies*, 2(3) (2008): 81-99.

—. Africa's Environmental Crisis: Unmapped Terrain, Existing Challenges and Possible Solutions. PhD Thesis. Binghamton University, State of New York, 2005.

Dodds, F. *The Earth Summit. A New Deal*. London: Earthscan, 2000.

Dogbe, K.B. "Sources of Cultural Knowledge, Objects, Sights And Sounds, Museums." In Cultural and Education: Report of National Conference on Culture (NCC) and Education at Elmina Beach Resort, 15-21 May 2005, edited by M.B. Abdallah 100-107. Accra: Sub-Saharan Publication,2009.

Donkor, B. N and Vloski, R. P. "A Review of the Forestry Sector in Ghana: A Louisiana Forest Products Development Centre", working paper No. 61, 2003.

Dopamu P.A. "Scientific Basis of African Magic and Medicine: The Yoruba Experience" In: *African Culture, Modern Science and Religious Thought* edited by P A. Dopamu. Ilorin: Decency Printers, 2003.

Dorm-Adzobu, C., Ampadu-Agyei, O. and Veit, P. G. *Religious Beliefs and Environmental Protection: The Malshegu Sacred Grove in Northern Ghana*. From the Ground Up Case Study No. 4, Centre for International Development and Environment, World Resources Institute, Washington, DC., 1991.

Dougherty, J.W.D. and C.M. Keller. "Taskonomy: A Practical Approach to Knowledge Structures." In *Directions in Cognitive Anthropology.* edited by J.W.D. Dougherty,161-174 Urbana: University of Illinois Press,1985.

Drazen, A. *Political Economy in Macroeconomics*. Princeton: Princeton University Press, 2000.

Driver, Harold Edson. "Culture Elements Distribution. Girls' Puberty Rites in Western North America" *University of California Anthropological Records* 6 (1941): 21-90.

Dunlap, R. E and Mertig, A. G. (eds.). *American Environmentalism: The U.S. Environmental Movement, 1970-1990*. Philadelphia, Washington, London: Taylor and Francis, 1991.

Durkheim, E., *The Elementary Forms of Religious life*: A new translation by C. Cosman. Oxford: Oxford University Press, 2008 (Re-used).

Dwivedi, O. P. "Satyagraha for Conservation: Awakening the Spirit of Hinduism." In *This Sacred Eearth:Religion, Nature, Environment.* edited by R. S. Gottlieb 151-163.London and New York: Routledge, 1996..

Dyasi, H. M. Culture and the Environment in Ghana. *Environmental Management 9(2)* (1985): 97-104.

Edjekumhene, I and Cobson-Cobbold ,J. C., KITE.2011. "Country Case Studies: Ghana." In *Low-Carbon Africa: Leapfrogging to Green Future Leapfrogging*, edited by A. Doig and M. Adow, Christian Aid, 53-56. Accessed June 24, 2012
http://www.christianaid.org.uk/images/LowCarbonAfrica.pdf.

Edusah, S.E. "Land Use, Land Allocation and the Environment in Kumasi and its Peri-Urban Communities of Ghana." The 12th EADI General Conference: Global Governance for Sustainable Development: The Need For Policy Coherence and New Partnership, Geneva, 24-28 June 2008.Accessed May 12, 2012
http://www.eadi.org/fileadmin/Documents/Events/General_Conference/2008/paper_Edusah_2.pdf

Eisenhardt, K. M. "Building Theories From Case Study Research." *Academy of Management Review*, 14(4) (1989): 532-550.

Ejizu C.I "Emergent Key Issues in the Study of African Traditional Religions.".Accessed February 4, 2011.
http://www.afrikaworld.net/afrel/ejizu.htm.

Ekem, John J. K. *Priesthood in Context*, Accra: SonLife Press, 2008.

Eliade., M. *The Sacred and the Profane: The Nature of Religion*. Harcourt: Brace and Inc., 1959.

Ellen, R.F. *Environment, Subsistence and Systems: The Ecology of small-scale Social Formations*. Cambridge: Cambridge University Press, 1982.

—. "What Black Elk left unsaid: On the Illusory Images of Green Primitivism." *Anthropology Today* 2 (6) (1986): 8-12.

—. "Rhetoric, Practice and Incentive in the face of Changing Times: A Case Study in Nuaulu Attitudes to Conservation and Deforestation."

In: *Environmentalism: The view from Anthropology*, edited by K. Milton London and New York: Routledge, 1993.

Ellen, R. F, Parkes, P., and Bicker, A. (eds.) *Indigenous Environmental Knowledge: Critical Anthropological Perspective*. London and New York: Routledge, 2003.

Ellis, Stephen and ter Haar Gerrie.*Worlds of Power: Religious Thought and Political Practice in Africa*. London: Hurst and Company, 2004.

Ekeke, E. C.; Ekeopara, C. A .2010. "Phenomenological Approach to the Study of Religion: A Historical Perspective." *European Journal of Scientific Research*; 44 (2) (2010):266-274.

Emmanuel Presbyterian Church, Berekum. Centenary Celebration and Thanksgiving service, 2006.

Ephirim-Donkor, A. African Religion Defined: A Systematic Study of Ancestor Worship among the Akan. Lanham, Boulder, New York, Toronto, Plymouth, UK: University of America Inc. 2010.

Evans-Pritchard, E. E. *Theories of Primitive Religion*, Oxford: Oxford University Press, 1987[reprinted].

Fairhead, J. and Leach, M. *Misreading the African Landscape: Society and Ecology in a Forest-Savannah Mosaic.* Cambridge: Cambridge University Press, 1996.

Fallers, L. *The King's Men: Leadership and Status in Buganda on the Eve of Independence*. London: Oxford University Press, 1964.

Falconer, Julia. "Non-Timber Forest Products in Ghana". London: Overseas Development Administration (ODA) Main Report, 1992.

Fargey, P.J. Assessment of the Conservation status of the Buabeng-Fiema Monkey Sanctuary. Report to the Flora and Fauna Preservation Society, 1991.

Faugier, J and Sargeant M. "Sampling Hard to Reach Populations." *Journal of Advanced Nursing*, 26(1997): 790-797.

Ferguson, Michael A. D and Messier F. "Collection and Analysis of Traditional Ecological Knowledge about Population of Arctic Tundra Caribou, Arctic 50 (1) (1997):.17-28.

Fernández, W. D. ."The Grounded Theory Method and case study data in IS Research: Issues and Design. Information Systems Foundations Workshop:Constructing and Criticising." Accessed March11, 2012. http://epress.anu.edu.au/info_systems/mobile_devices/ch05.html, 2004.

Fetzner, W. J. Jnr. Ghana Butterfly y Project. Accessed June 6,2012. Available from: http://iz.carnegiemnh.org/GhanaBFly/Background.asp, 2007.

File: Side-Striped Jackal.JPG. Accessed June 12, 2011

http://en.wikipedia.org/wiki/File:Side-.

Fisher, B.L "Insect Behaviour and Ecology in Conservation: Preserving Functional Species Interactions." *Ann Ent. Soc Am* 91 (1998):155-158.

Fitzgerald, T. *The Ideology of Religious Studies*. New York: Oxford University Press., 2000

Flood, G. D. *Beyond phenomenology: Rethinking the Study of Religion.* London: Cassell, 1999.

Flolu, J. "Some Approaches to Teaching Culture." *In*: Cultural and Education: Report of National Conference on Culture (NCC) and education at Elmina Beach Resort,May 15-21, 2005, edited by M.B. Abdallah, 59-70 Accra: Sub-Saharan Publication, 2009.

Food and Agriculture Organisation (FAO) -Twenty-third Tegional Conference for Africa. 2004. The Bushmeat Crisis in Africa: Conciliating Food Security and Biodiversity Conservation in the Continent, 1-5 March 2004, Johannesburg, South Africa.

Fomin, E. S. D. "Royal Residences and Sacred Forests in Western Cameroon: The Intersection of Secular and Spiritual Authority." *Journal for the study of Religion, Nature and Culture* 2(3 (2008):391-407.

Forestry Inventory Management Project (FIMP) 1994 Report.

Forestry Outlook Study for Africa (FOSA) and the Ghanaian Ministry of Lands and Forestry. 2002. Bad Management in Ghanaian Forestry Sector. Based on FOSA and Afrol Archives. [Accessed May 30, 2012. http://www.afrol.com/News2002/gha007_forestry.htm.

Fortes, M. "The Dynamics of Clanship among the Tallensi." In *Land Tenure and Agrarian reform in Africa and the near east: An Annotated Bibliography.* edited by T.J. Anderson , 67, Boston, Mass: G.K. Hall and Co. 1945.

Frazer, J. G. *Totemism and Exogamy*, 4 vols. London: Macmillan, 1910.

Freud, Sigmund. *Totem and Taboo: Resemblances between the mental lives of Savages and Neurotics*: Imago. Original title: *Totem und tabu: Einige ubereinstimmungen im seelenleben der wilden und der neurotiker*, 1912-13.

Gathogo,J.M. "The Reason for Studying African Religion in Post-Colonial Africa." *Currents in Theology and Mission*, 36(2) (2009):108-117.

—. The Relevance and Influence of African Religion in post-Apartheid South Africa and beyond part 2, *Churchman*, 121 (3) (2007b):249-256.

Geertz, C.*The Interpretation of Cultures: Selected Essays*. USA: Basic Books, 1973.

Ghana. Country Environmental Profile., 2006 Accessed June 4, 2013]. :http://ec.europa.eu/development/icenter/repository/Ghana%20_CEP_2006.pdf.

Ghana Districts—A Repository of all Districts in theRrepublic of Ghana. ghanadistricts.com. Accessed October 10, 2010. http://www.ghanadistricts.com/districts/?r=10&_=34&sa=4867

Ghana News Agency (GNA). "State of Environment Report Launched. 2004.".Accessed 12 February 12, 2012. http://www.modernghana.com/news/92281/1/state-of-environment-report-launched.html

—. Ghana ranked 48th in Africa for poor sanitation, 2008. Accessed January 2, 2012. http://www.ghanaweb.com/GhanaHomePage/NewsArchive/artikel.php?ID=148014

—. Annual cost of Environmental Degradation to Ghana: $1.2BN. 2007. Accessed May 13, 2012. http://www.modernghana.com/news/122558/1/annual-cost-of-environmental-degradation-to-ghana-.html

Ghanaian Times, August 29, 2012 edn.

Gilgun. J.F. "Reflexivity and Qualitative Research." *Current Issues in Qualitative Research,* 1(2) (2010):1-8.

Githitho, A.N. "The Sacred Mijikenda Kaya Forests of Coastal Kenya and Biodiversity Conservation." In *The Importance of Sacred Natural SitesfFor Biodiversity Conservation. Proceedings of the International Workshop held in Kunming and Xishuangbanna Biosphere Reserve, People's Republic of China, 17–20 February 200,* edited by C. Lee and T. Schaaf, 27-35. Paris: United Nations Educational, Scientific and Cultural Organization (UNESCO),.Accessed June 20, 2013] Available from: http://unesdoc.unesco.org/images/0013/001333/133358e.pdf

Glaser, B. G. Theoretical *Sensitivity: Advances in the Methodology of Grounded Theory*. Mill Valley, CA: Sociology Press, 1978.

—*The Grounded Theory Perspective: Conceptualization Contrasted with Description*. Mill Valley, CA: Sociology Press, 2001.

Goldstein, M., & Udry, C. *The Profits of Power: Land Rights and Agricultural Investment in Ghana.* The World Bank, 2008.

Gonese, C. 1999. Culture and Cosmovision of Traditional Institutions in Zimbabwe. In *Food for Thought: Ancient Visions and New Experiments of Rural People*, edited by W. Haverkort, B. and W. Hiemstra 155-162. Leusden: COMPAS, London: Zed Books, 1999.

Gottlieb, R.S. A Greener Faith: Religious Environmentalism and Our Planet's Future. Oxford: Oxford Univ. Press, 2009

Gordon, C. "Sacred Groves and Conservation in Ghana. *Newsletter of the IUCN, African Reptile and Amphibian Specialist Group* 1(1992):.3-4.

Graham, T. C. "Using Reason for Living to Connect to American Indian Healing Traditions." *Journal of Sociology and Social Welfare*, XXIX (1) (2002): 55-75.

Groves, C. P.*The Planting of Christianity in Africa, 4 Vols*. London: Lutterworth, 1948-58.

Gyasi, E. A. and. Uitto, J. I (eds.).*Environment, Biodiversity and Agricultural Change in West Africa: Perspectives from Ghana*, Tokyo: United Nations University Press, 1997.

Gyekye, K.*African Cultural Values: An Introduction, Accra*: Sankofa Publishing Company, 1996.

Hagan, G.P. "Traditional Laws and Methods of Conservation and Sustainable use of Biodiversity.". In . *Proceedings of the third UNESCO MAB Regional Seminar on biodiversity conservation and sustainable development in Anglophone Africa (BRAAF), Cape Coast, 9-12th March 1997.* edited by D.S. Amlalo, L.D. Atsiatorme and C. Fiati. Accra: Environmental Protection Agency (EPA), 1998.

Hannibal, L. A. Environmental Sociology. A Social Constructionist Perspective. London: Routledge, 1995.

Hargrove, E. C. (ed.). *Religion and Environmental Crisis*. Athens, GA: University of Georgia Press, 1986.

Harrison, Peter. 'Subduing the Earth: Genesis1, Early Modern Science, and the Exploitation of Nature.' *The Journal of Religion* 79 (1)(1999):.86-109.

Hart, Michael. Anthony. 'ndigenous Worldviews, Knowledge, and Research: The Development of an Indigenous Research Paradigm.' *Indigenous Journal for Social Work*, 1 (1) (2010):1-16.

Hayford E. K., Amin, A. OsaE, *E.K.* and Kutu, J. Impact of Gold Mining on sSil and some Staple Foods Collected from selected Mining Communities in and around Tarkwa-Prestea area, *West African Journal of Applied Ecology,* (14) (2008): 1-12.

Haynes, J. *Religion and Development: Conflict or Cooperation*? New York: Palgrave, Macmillan, 2007.

Hays, S.P. 'eauty, Health and Permanence: Environmental Politics the United States, 1955-1985.' New York: Cambridge University Press, 1987.

Heckathorn, D. D. 2002. "Respondent-Driven Sampling II: Deriving Valid Estimates from Chain-referral Samples of HiddenPpopulations". *Social Problems* 49 (1) (2002):11–34.

Helminiak, Daniel A. *The Human Core of Spirituality*. Albany: State University of New York Press, 1996.

Holloway, I. and Wheeler, S. *Qualitative Research for Nurses*. Oxford, Cambridge, Mass: Blackwell Science, 1996.

Holy Family Hospital, Berekum. "Pollution-Related Diseases Records, 2007-2011." 2012.

Hussain .S. "Nature and Human Nature: Conservation, Values And Snow Leopard." Prepared for the Snow Leopard Survival Strategy Workshop, Seattle, USA May 2002). Accessed December 29, 2010. http://www.snowleopardnetwork.org/bibliography/hsnh02.pdf, 2002

Husserl, Edmund. *Ideas. General Introduction to Pure Phenomenology*, translated by W.R. B. Gibson. London: Allan and Unwin Ltd., 1931.

Idowu, E. Bolaji *African Tradition Religion; A Definition.* London SCM Press Ltd., 1973.

Ikenga-Metuh, Emefie . The Gods in Retreat: Continuity and Change in African Religion. Enugu, Nigeria: Fourth Dimension Publishers Co., 2002.

—. *Comparative studies of African Traditional Religions*. Onitsha, Nigeria: IMICO Publishers, 1987.

—. *African Religions in Western Conceptual Schemes*. Ibadan: Pastoral Institute, 1985.

Ingold, Tim. 'Culture and the perception of the environment. In *Bush Base: Forest Farm*, edited by E.Croll, and D. Parkin 39-56. London: Routledge, 1992.

INCN (International Union for Conservation of Nature). "The IUCN Red List of Threatened Species."Accessed March 20, 2012 http://www.iucnredlist.org/documents/redlist_website_users_guide.pdf

Jackal, Jerry. "The Spirits of Conservation: Ecology, Christianity, and Resource Management in Highlands Papua New Guinea", *Journal for Study of Religion, Nature and Culture,* 4 (1) (2010):.24-47.

Jehu-Beaulaton, D. "Sacred forests and the global challenge of biodiversity conservation: The case of Benin and Togo". *Journal for the Study of Religion, Nature and Culture,* 2(3) (2008):.351-372.

Johnson, Philip. N. and Butigan, Ken. (eds.). *Cry of the environment: rebuilding the Christian creation tradition*. Santa Fe, NM: Bear and Company, 1984.

Johnson, William Theodore. "The Bible on environment conservation: A 21st century Perspective." *Quodlibet Journal* 2 (4). Accessed 29 December 2010. http:www.quodlibet.net/articles/Johnson-environment.shtml.

Jones, M. and Alony, Irit. "Guiding the use of Grounded Theory in Doctoral Studies-An example from the Australian Film Industry." *International Journal of Doctoral Studies,* 6 (2011):.99- 113.

Jorgensen, D. L. *Participant Observation: A Methodology for Human Studies*, Newbury Park, CA: Sage Publications, 1989.

Kalu,.U Obgu "Ancestral Spirituality and Society in Africa." In .*African Spirituality: Forms, Meanings and Experiences*. Edited by J. K Olupona, 54-84. New York: The Crossroad Publishing Company, 2000.

Kamwaria, Alex and Katola, Michael The role of African traditional religion, culture and worldview in the context of post- war healing among the Dinka community of southern Sudan. *International Journal of Humanities and Social Science,*2 (21) (2002): 48-55.

Kendie, S. B. and Guri, B. Indigenous Institutions, Governance and Development: Community Mobilisation and Natural Resource Management in Ghana in Endogenous Development and Bio-Cultural Diversity, (2006) 332-349.Accessed 07-May 7,2011 http://www.bioculturaldiversity.net/Downloads/Papers%20participants /Kendie_Guri.pdf.

King, Ursula. *Religion and Gender*. Oxford: Blackwell, 1995.

Kingdom J. *Island Africa: The Evolution of Africa's Rare Animals and Plants.* Princeton, N. J: Princeton University Press, 1989.

Kingsley, D. R.*Ecology and Religion: Ecological Spirituality in cross Cultural Perspective*. Englewood Cliff NJ: Prentice-Hall, 1995.

KOB Consulting Ltd. Civil Service Performance Improvement Programme: Beneficiary Survey for Berekum District Assembly, December 1988.

Kothari, A. Traditional Knowledge and Sustainable Development, *International Institute for Sustainable Development (IISD)* (2007). Available at: http://www.iisd.org.

Kraft, Charles. H. "Culture, Worldview and Contextualisation." . In *Perspectives on the World Christian Movement*. 3d. Edn. edited by R. D. Winter and S.C. Hawthorne, 84-391.Pasadena: William Carey Library, 1999.

Kumar De A. and Kumar DE A. *Environment and Ecology*. New Dheli New Age International Publishers, 2009.

Laird, A. S. "Trees, Forests and Sacred Groves." Accessed August 23, 2012. http://www.agroforestry.net/overstory/overstory93.html, (not dated).

Learning About Forest Products (LEAF) . "Forest Myths. "Accessed January 20, 2012. http://leaf-international.org/article.cfm?ID_art=23&ID_kanal=5, 2009

Lebbie, A. and Guries, R. P. "The Role of Sacred Groves in Biodiversity in Sierra Leone." In *African sacred groves: Ecological dynamics and social change*. edited by M. J. Sheridan, and C. Nyamweru, 42-61. Oxford: James Curry; Athens, OH: Ohio University Press, 2008.

Leopold. A. *A sand county almanac: With Essays on Conservation from Round River*. New York, Sierra Club and Ballantine, 1986 [1949]. .

Leslie, E. S (contributing author); Casagrande . D (topic editor) . "Religion, Nature and Environmentalism." In *Encyclopeadia of Earth* edited by J.C. Cutler. Washington, D.C. Environmental Information Coalition, National Council for Science and the Environment, 2007.

Lester, S. "An Introduction to Phenomenological Research. "Taunton UK, Stan Lester Developments.Accessed October 10, 2013. http://www.sld.demon.co.uk/resmethy.pdf., 1999.

Lewis, J. L. and Sheppard S. R. J. "Culture and Communication: Can Landscape Visualization Improve Forest Management Consultation with Indigenous Communities?" *Landscape and Urban Planning* 77 (2006): 291–313.

Lewis, I. M. (ed.) "Islam in Tropical Africa: Studies presented and discussed at the 5th International Africa Seminar, Ahmadu Bello University Zaria, January 1964." Oxford: Oxford University Press, 1967.

Little Bear. L. "Jagged Worldviews Collide." In *Reclaiming Indigenous Voice and Vision* edited by M. Battiste Vancouver, British Columbia, Canada: UBC Press (2000):.77-85.

Little, P.D. and Brokensha, D. Local Institutions, Tenure and Resource Management. In *Conservation in Africa: People, Policies and Practice* edited by D. Anderson and R. Grove, 193-209 .Cambridge and New York: Cambridge University Press.

Lugo A. E "Management of Tropical Biodiversity." *Ecological Applications*, 5 (1995):. 956-961.

Maathai, Wangari. "Replenishing-the-Earth." Accessed, January 29, 2012: http://www.yesmagazine.org/happiness/wangari-maathaispiritual-environmentalism-healing-ourselves-by-replenishing-the-earth., 2011

Mafeje, A. *In Search of an Alternative: A Collection of Essays on Revolutionary Theory and Politics*. Harare, Zimbabwe: SAPES Books, 1992.

Magesa, Laurenti. *African Religion: The Moral Traditions of Abundant Life*. Mary Knoll: Orbis,1997.

Martin, P. Y. and Turner, B. A. "Grounded Theory and Organizational Research." *The Journal of Applied Behavioral Science*, 22(2) (1986):.141-157.

Mays, N. and Pope, C. "Quality in Qualitative Health Rresearch." In *Qualitative Research in Health Care* (2nd edn.) edited by N. Mays and C. Pope, 89-102. London: BMJ Books, 2000

Mavetera, N., Kroeze, J.H. "Practical Considerations in GroundTd theory Research' Sprouts: Working Papers on Information Systems." 9(32). Accessed 25 June 25, 2013: http://sprouts.aisnet.org/9-32, 2009.

Mawere. Mayaradzi. *Environmental Conservation through Ubuntu and other Emerging Perspectives. Mankon.* Bamenda: Langaa Research and Publishing CIG, 2014a.

—. "A Critical Investigation of Environmental Malpractices in Mozambique: A Case Study of Xai-Xai Cmmunal Area, Gaza Province, Mozambique." *International Research Journals,* 2 (2) (2011): 874-883.

—. *Culture, Indigenous Knowledge and Development in Africa: Reviving Interconnections for Sustainable Development,* Bamenda, Cameroon, Langaa RPCIG, 2014b.

—. *African belief and knowledge systems: A critical perspective,* Langaa RPCIG, Bamenda, 2011.

—. *The struggle of African indigenous knowledge systems in an age of globalisation: A case for children's traditional games in south-eastern Zimbabwe*. Bamenda: Langaa RPCIG, 2012.

Mazrui, A. Ali. *The Africans: A Triple Heritage.* Boston and Toronto: Little Brown and Company, 1986.

Mbiti, John S *Introduction to African Religion 2nd ed.* Nairobi, Kampala, Dar es Salam: African Educational Publishers, 1991.

—. *African Religions and Philosophy*. London: Heinemann,1969.

—. *Akamba Stories.* Oxford, Clarendon Press, 1966.

Mbiti, John. S. and Burleson, B. W. The Dialogue of an African Theologian with African Religion. Ann Arbor, Michigan: University Microfilms International, 1986.

Mccann, Terence and Clark, E. "Grounded Theory in Nursing Research: Part 2 Critique." *Nurse Researcher,* 11 (2) (2003):19-28.

McCaskie, T.C. *State and Society in Pre-Colonial Asante.* Cambridge: Cambridge University Press, 2002 [First paper back edition).

Mccutcheon R. T *Manufacturing Religion: The Discourse on sui generis Religion and the Politics of Nostalgia*. New York and Oxford: Oxford University Press, 1997.

—. "The Insider/Outsider Problem in the Study of Religion: A Reader." London and New York: Cassell, 1999.

Mcgregor, D. "Traditional Ecological Knowledge and Sustainable Development: Toward Co-Existence." In.*The Way of Development:*

IndigenousPpeoples, Civil Cociety and the Environment, edited by M. Blaser and H. Feit. 72-91. New York: Zed Books,2004

Mclennan, I. F. "The Worship of Animals and Plants." *Fortnightly Review,* 6 (1869-70): 407-427.

Mcleod, M. D. *The Ashanti*. British Museum Publication Ltd, 1981.

Mcneeley, J.A. "Achieving Financial Sustainability in Biodiversity Conservation programme." Gland, Switzerland: IUCN, 1999. .

Menzies, C.R. (ed.).*Traditional Ecological Knowledge and Natural Resource Management.* Lincoln and London: University of Nebraska, 2006.

Meyerowitz, E. *The Akan of Ghana: Their Ancient Beliefs*. 24 Russell Square London: Faber and Faber Limited, 1958.

Mkhize, N. "Psychology: An African perspective." In *Self, Community and Psychology* edited by K. Ratele, N. Duncan, D. Hook, N. Mkhize, P. Kiguwa, and A.Collins. Republic of South Africa,. Lansdwone: UCT Press, 2004.,.

Migdal, J.S. *Peasants, Politics and Revolution: Pressures toward Political and SocialCchange in the Third World*. Princeton: Princeton University Press, 1974.

Millar, D. "Traditional African Worldviews form a Cosmovision Perspective." In *Food for Thought: Ancient Visions and new Experiments of Rural People*, edited by B. Haverkort and W.Hiemstra, 155-162 Leusden: COMPAS, London: Zed Books,1999..

Millar, D, Ausigah. A. A. and Berinyuu, A. The Chief, the Forester and the Fireman: Workshop Proceedings, University of Development Studies, Bolgatanga, Ghana, 2004.

Milton, K. *Environmentalism and Cultural Theory: Exploring the role of Anthropology in Environmental Discourse*. New York and London: Routledge, 1996.

—. "Nature and the Environment in Indigenous and Traditional Cultures." In *Spirit of the Environment: Religion, Values and Environmental Concern.* Edited by E. David and, J. A. Palmer, 86-99. London and New York: Routledge, 1998.

Ministry of Environment and Science. National Biodiversity Strategy, Accra, Ghana, 2002.

Ministry of Land and Forestry (MLF). Forest and Wildlife Policy (FWP), 1994.

Ministry of Land and Forestry (MLF). Forestry Sector Development Maste Plan (FSDMP). Accra. Government of Ghana, 1996.

Mitchell, C. J. Case and Situation Analysis. *The Sociological Review* New Series, 31(2) (1983):187-211.

Morgan, D. L. "Qualitative Content Analysis: A Guide to Paths not taken." *Qualitative Health Research*, 3 (1993):112-121.

Morris, M. G., Collins, N. M., Vane-Wright, R. I and Waage, J. "The Utilisation and Value of non-Domesticated Insects." In *The Conservation of Insects and their Habitats,* edited by N. M. Collins and J .A. Thomas, 319-347. Londres Academic Press limited, 1991.

Morse, J. M. "Qualitative Generalisability." *Quality Health Research*, 9(1999):.5-6.

Mpara, J. "Indigenous Knowledge Systems in Zimbabwe: Juxtaposing Post-Colonial Theory." *Journal of Pan African Studies*, 3 (1) (2009): 139-155.

Mugambi, J.N.K. *Religion and Social Construction of Reality*. Nairobi: Nairobi University Press, 1996.

Mufeme, E. "Land: Breaking Bonds and Cementing ties." *Echoes. land and spirituality in Africa series.* World Council of Churches (WCC).Accessed 23 June 23, 2012. http://www.wcc-coe.org/wcc/what/jpc/echoes-16-05.html, 1999.

Mukamuri, B. "Making Sense of Social Forestry: A Political and Contextual Study of Forestry Practices in South-Central Zimbabwe." PhD Dissertation, University of Tampere, Finland, 1995.

Myers, M. "Qualitative Research and the Generalisability Question: Standing firm with Proteus." *The Qualitative Report*, 4(3/4) (2000). Accessed June 10, 2011 http://www.nova.edu/ssss/QR/QR4-3/myers.html.

Naess, Arne. "The Shallow and the Deep, Long-Range Ecology Movement" *Inquiry* 16 (1973): 95-100.

—. The Shallow and the Deep, Long-Range Ecology Movement.' *Inquiry* 16 (1973):.95-100.

Nana Oduro Kwarteng. "Poor sanitation mainly an attitudinal problem." *Ghana News Agency (GNA)* January, 16 2013 Edition.Accessed 14 June 14, 2013 http://ghananewsagency.org/social/poor-sanitation-mainly-an-attitudinal-problem-55036, 2013.

Nkansa-Kyremanteng, K. *Akan Heritage*. Accra: Sabewie Publishers, 1999.

Nazarea, V.D., Robert, E. R., Bontoyan, E and Flora, G. "Defining Indicators which make sense to Local People: Intra-Cultural Variation in Perceptions of Natural Resources" *Human Organization* 57(2) (1998):159-170.

Nel, P. J. "Morality and Religion in African Thought." *Acta Theologica*, 2, (2008):33-47.

Nkongmeneck B.A. Lowman M.D and Atwood J.T. 2002 "Epiphyte Diversity in Primary and Fragmented Forests of Cameroun, Central Africa: a Preliminary Survey" *Selbyana*, 23 (1) (2002):121-130.

Nsiah, K. M. "Culture and Climate Change Interface, with Particular Reference to Ghana" Accessed March 5 2012 http://www.openmeeting2009.org/pdf_files/Pdf%20papers/Michael%2 0kofi% 20Nsiah.p df, 2009.

Ntiamoa-Baidu, Yaa. "onservation of Coastal Lagoons in Ghana: the Traditional Approach". *Landscape and Urban Planning,*. Amsterdam.Elsevier Science Publishers B.V, 20, (1991):41-46.

—. "Indigenous Versus Introduced Biodiversity Conservation Strategies: The Case of Protected Areas Systems in Ghana." *African Biodiversity Series*, no. 1, 1995): 1-11 Biodiversity support programme.

—. "Indigenous beliefs and biodiversity conservation: The Effects of Sacred Groves, Taboos and Totems in Ghana for Habitat and Species Conservation," *Journal for the Study of Religion, Nature and Culture 2(3)* (2008):309-326.

Ntiamoa-Baidu, Y., Soutter, R., Smith, J and Devendra.R "Recognising the Contribution of Sacred Natural Sites for Biodiversity Conservation." A paper presented at the World Parks Congress in Durban, South Africa, September, 2003.

Nugteren, A. "From Cosmos to Commodity..and back: A Critic of Hindu Environmental Rhetoric in Educational Programs". In *Religion And Sustainable Development Opportunities and Challenges for Higher Education*.edited by C. De pater and I. Dankelman,159-168 Berlin: Lit Verlag, 2009.

Nukunya, G. K. "Christianity, Western Education and Social Change: An overview." In *Christian Missionarism and the Alienation of the African Mind* edited by S ,Ramadan. Belhag and A .E'.K. YassiN, 87-98. Tripoli: African Society of Social Sciences, 1986.

Nwosu, P.U "The Role of Okonko society in Preserving Igbo Environment." *Journal of Human Ecology,* 31(1),(2010): 59-64.

Obeng. E.A. "The Religious Experience And The Environment: Implication For Religious Understanding." In *Journal of African religion and philosophy*, 2 (1) edited by L. Njinya-Mujinya, M.J.K. Muranga,O. Ndoleriire and P. Habonmugisha .121-126, 1991.

Obiri, S. Determination of Heavy Metals in Boreholes in Dumasi in the Wassa West district of Western Region of the Republic of Ghana. Environmental Monitoring and Assessment 130 (2007): 455-463.

Odotei, Irene K. *Chieftaincy in Ghana*. A paper presented to German students on the Go Africa, Go Germany Programme, at Noda Hotel, Fumesua – Kumasi on 20th March, 2010.

Odotei, Irene K., and Awedoba, Albert K. (eds.*) Chieftaincy in Ghana: Culture, Governance and Development.* Accra: Sub-Saharan Publishers, 2006.

Ofori. S. C. "Environmental Impact Assessment in Ghana: Current Administration and Procedures: Towards Appropriate Methodology." *Environmentalist.* 11 (1) (1991): 45-54.

Ojomo, P. A "An African Understanding of Environmental Ethics: Thought and Practice." *A Journal of Philosophical Association of Kenya, New Series*, 2 (2) (2010): 49-63.

Olivier, S.M. Information technology research: A *Practical Guide For Computer Science And Informatics,* 2nd ed. Van Schaik, Pretoria, South Africa, 2004.

O'leary, Z. *The Essential Guide to DoingRresearch*. Los Angeles, London, New Dheli, Singapore: Sage Publications, 2004.

Olupona, J.K (ed.). *African Spirituality: Forms,Mmeanings and Experiences*. New York: The Crossroad Publishing Company, 2000.

Olupona, J.K (ed.). *African Traditional Religion in Contemporary Society*.St. Paul, Minneapolis: Paragon House, 1991.

Omari, C. K. "Traditional African Land Ethics." In: *Ethics of Environment and Development: Global Challenge, International Response,* edited by J Ronald J., Engel, and E. J.oan Gibb, 167-175 Tucson, Arizona: University of Arizona Press, 1990.

Onyinah, O. Akan Witchcraft and the Concept of Exorcism in the Church of Pentecost. PhD. Thesis, the University of Birmingham, 2002.

Opoku-Ankoma, Y., Amevenku, F. K. Y. and obuobie, E."Research Findings in Water Bodies:the Human and the Cultural Factors and Their Impact on The Environment,. Report of Nation Conference on Water." 24-26 February at Continental Hotel, Akosombo,(2010):20-26.

Opoku, K. A. *West African Traditional Religion*. Accra, Bangkok, Hong Kong: Fep International Private Ltd., 1978.

Osei, Joseph. "African Traditional Taboos and Development: Insight from A Philosophical Analysis of Akan and Ewe Taboos.". Accessed August 11 2010: http://www.jsd-africa.com/Jsda/Fall2006/PDF/Arc_the%20Value%20 of%20Arican%20Taboos.pdf

Osuntokun, A. *Environmental Problems of the Niger-Delta*. Ibadan: Davidson Press, 2001.

Owusu, B. Chieftaincy and traditional taboo: An empirical approach. In *Chieftaincy in Ghana: Culture, Governance and Development* edited by I.K. Odotei, and K. A. Awedoba , 213-230 Accra: Sub-Saharan Publishers, 2006.

Pals, D. L. *Eight Theories of Religion* 2nd ed. New York: Oxford University Press, 2006.

Palmer, J A. "Destruction of the Rain Forest: Principles and Practices?" In *The Environmental Question: Ethics And Global Issues,* edited by D.E Cooper and J.A. Palmer,81-83. London and New York: Routledge, 1992.

Pandey, D.N. 2003. "Cultural Resources for Conservation Science." *Conservation Biology*. 17 (2)(2003):633-635.

Park, Chris .*The Environment: Principles and Applications 2nd*. London and New York: Routledge, 2007.

—. A *Dictionary of Environment and Conservation.* Oxford: Oxford University Press, 2007.

Parkin, D. "Politics Of Ritual Syncretism: Islam among the non-Muslim Giriama of Kenya. Africa" *Journal of the International African Institute*. 40(3) (1970): 217-233 .

Parrinder, E. G. *West African Religions: A Study of the Beliefs and Practices of Akan, Ewe, Yoruba, Ibo and Kindred Peoples.* London: The Epworth Press, 1961.

—. *African Traditional Religion, 3rd ed*. London: Sheldon Press, 1974.

Patton, M.Q. *Qualitative evaluation and research methods 2nd edn.* Newbury Park, CA: Sage, 1990.

Pichette, E. F., Garrett, M. T., Kosciulek, J. F. and Rosenthal, D. A. "Cultural identification of American Indians and it's [sic] Impact on Rehabilitation Services." *Journal of Rehabilitation*, (3)(1999): 3-10.

Pike, K. L. *Language in Relation to a Unified Theory of the structure of Human Behaviour 2nd. ed*. The Hague: Mouton, [1954, 1955, 1960] 1967.

Pobee, John.S. *Towards an African Theology*. Nashville: Abingdon, 1979.

Polit, D.F., Hungler, B.P. *Essentials of Nursing Research Methods, Appraisal, and Utilization 3rd ed*. Philadelphia: J.B. Lippincott company, 1993.

Postma, D. W *Why Care For Nature? In Search Of An Ethical Framework For Environmental Responsibility And Education*. Dordrecht, The Netherlands: Springer, 2006.

Porteos, A. *Dictionary of environmental science and technology*, 3rd ed. Chester: Willey, 2000.

Prime, Ranchor. *Hinduism and Ecology: Seeds of Truth*. London and New York:Cassell, 1992.

Ratcliff, D. Qualitative Research. Part five: Data Analysis.Accessed May 4, 2013 http://www.don.ratcliff.net/qual/expq5.html., 2002

Rattray, Robert S. *Ashanti.* London: Oxford University Press, 1959.

—. *Religion and Art in Ashanti*. London: Oxford University Press, 1923.

—. *Akan–Ashanti Folktales.* Oxford, Clarendon Press, 1930.

Ray, Benjamin. C. *African Religions: Symbols, Rituals and Community*. Englewood Cliffs, New Jersey: Prentice Hall Inc, 1976.

Riitters K, J, Wickham., R., O'Neil, B. J. and. Smith, E. 2000. "Global Scale Patterns of Forest Fragmentation.", *Conservative Ecology*, 4 (2000):.1-29.

Rockefeller, S. C. and John, C. E.(eds.) *Spirit And Nature: Why the Environment is a Religious Issue: An Interfaith Dialogue.* Boston. Beacon Press, 1992.

Rolston, III, Holms. *Environmental Ethics: Duties to and Values in the Natural World*. Philadelphia: Temple University Press, 1988.

Rose, D., James, D. and Watson, C. *Indigenous Kinship with the Natural World in New South Wale. New South Wales National Park and Wildlife Services* (NSW), 2003.

Rydving, H. and Geertz, Clifford. "Suggestions on Participatory bservation as Method of Data Collection." In *Research methods in Cultural Anthropology*, edited by B. Russell. Newsbury Park, California: Sage Publications, 1998.

Saint Theresa's Catholic Church. Berekum, Historical Records, 2009.

Sandelowski. M. Focus on Research Methods whatever happened to Qualitative Description? *Research in Nursing and Health* , 23(2000):334-340.

Sarbah, C. J. E. 2010. A critical study of Christian-Muslim relations in the central region of Ghana with special reference to traditional Akan values Ph.D. thesis, University of Birmingham.

Sarfo-Mensah, Paul., Oduro, William. and Anto -Fredua, E. 2010. Traditional Representations of the Natural Environment and Biodiversity Conservation: Sacred Groves in Ghana. Fondazione Eni Enrico Mattei Working Papers Accessed 20 August, 2011: http://www.bepress.com/feem/papers398., 2010.

Sarfo-Mensah, Paul. and Oduro, William. 2007. *Traditional Natural Resources Management Practices and Biodiversity Conservation in Ghana: A Review of Local Concepts and Issues on Change and Sustainability.* Accessed 20 Augst, 20, 2011].

http://www.feem.it/Feem/Pub/Publications/WPapers/default.htm,, 2007.

Sarfo-Mensah, Paul. The Spirituality of Forests And Conservation: The Dynamics of Change and Sustainability of Scared Groves in the Transitional Zone of Ghana .PhD. Thesis. University of Greenwich, UK, 2001.

Sarpong, P.K. *Girls nubility Rites in Ashanti*, Accra: Ghana Publishing Cooperation ,1977.

—. *Ghana in retrospect: Some aspects of Ghanaian culture*. Tema: Ghana Publishing Corporation, 1974.

—. Can Christianity Dialogue with African Traditional Religion?Accessed May 22, 2013] Available from: http://www.afrikaworld.net/afrel/sarpong.html (not dated).

Schaat, T. "UNESCO's Experience with the Protection of Sacred Natural Sites for Biodiversity Conservation." In *Importance of Sacred Natural Sites for Biodiversity Conservation. China,* 17-20 February 2003 edited by C. Lee.13-20. Paris, UNESCO,pp.13-20.

Schneider, D. K. "Methodology Tutorial-Qualitative Data Analysis" Accessed May 9, 2013. http://edutechwiki.unige.ch/en/Methodology_tutorial_-qualitative_data_analysis,2008.

Schoffeleers, J. M. (ed.). *Guardians of the Land: Essays on Central African Territorial Cults.* Harare, Zimbabwe: Mambo Press, 1978.

Schulkin, J., and D. Sarokin. "The Culture of Environmental Decision-Making." *Environmentalist* 16(1996): 283-289.

Sessions, G. (ed.) 1995. *Deep ecology for the twenty-first century* .Boston:Shambhala.

Sheridan, M. J., and C. Nyamweru (eds.). *African Sacred Groves: Ecological Dynamics And Social Change*. Oxford: James Currey; Athens, OH: Ohio University Press, 2008.

Shorter, A. *African Christian Theology*. London: Geoffrey Champmans, 1975.

Sinclair F. L., and Walker D. H."A Utilitarian Approach to the Incorporation of Local Knowledge in Agroforestry Research and Extension." In *Agroforestry in Sustainable Agricultural Systems* edited by L. E Buck, J. P Lassoie and E. C. M. Fernandes, 245-275. Florida: CRC Press, 1999.

Sinha, Subir, Shubhra Gururani and Brian Greenberg "The new traditionalist discourse of Indian environmentalism." *The Journal of Peasant Studies*.24 (3) (1997): 65-99.

Shiva, V. *Staying Alive: Women, Ecology and Survival in India*. New Delhi:Kali for Women, 1998.

Slikkerveer, L. Jan. "Ethnoscience, 'TEK' and its Application to Conservation." In *Cultural and Spiritual Values of Biodiversity* edited by D.A. Posey , 160-180.UNEP, Nairobi, Kenya, 1999.

Smart, N. *The religious experience of mankind*. New York: Charles Scribner's sons, 1969

—. *The phenomenon of religion.* London: Mowbrays, 1978.

—. *Concept and Empathy: Essays in the Study Of Religion*. Basingstoke MacMillan, 1986.

—. *Dimension of the sacred: An anatomy of the world`s beliefs*. London: Harper Collins, 1986.

Smith-Asante. E. 2002. "Totemism Under Threat..Our Totems Get Summarily Executed." *The Independent* .Accessed December 23, 2010 http://www.biodiversityreporting.org/article.sub?docId=757&c=Ghana &cRef=Ghana&ye ar=2003&date=November+2002.

Smith, E, (ed.). *African Ideas of God*. Edinburgh House Press, 1950.

Smith, N. *The Presbyterian Church of Ghana, 1835-1960: A younger Church in a changing society.* Accra: Ghana Universities Press, 1986.

Smith, E. A., and. Wishnie, M. "Conservation and Subsistence in Small-Scale Societies." *Annual Review of Anthropology* 29(2000):493-524.

Shoko, T. 2012. Teaching African Traditional Religion at the University of Zimbabwe. In:*African traditions in the study of religion in Africa*, edited by A, Adogame,.E, Chitando and B. Bayete, 52-65 England, USA:Ashgate, 2012.

Steedman, P.H. 1991: "On The Relations Between Seeing, Interpreting and Knowing." In *Research and Reflexivity* edited byF.Steier, 53-62 . London: Sage, 1991.

Storey, J. *Cultural Theory and Popular Culture: An Introduction* 3rd edn. London, New York: Prentice Hall, 2001.

Strehlow, T. G.H. "Geography and Totemic Landscape of in Central Australia: A Functional Study." In. *Australian Aboriginal Anthropology* edited by R. Berndt. Canbella: Autralian Institute of Australian Studies (1970): 92-140.

Sutcliffe, Steven. *Religion: Empirical Studies: A collection to mark the 50th Anniversary of the British Association for the Study of Religion.* Aldershot: Ashgate, 2004.

Sutton ,M. Q. and Anderson ,E .N. *Introduction to Cultural Ecology* 2nd edn. Lanham. New York, Toronto, Plymouth , UK: Altamira Press, 2010.

Swensson, John. "Bushmeat Trade in Techiman, Ghana, West Africa". 2005. Undergraduate Thesis, Uppsala University, Sweden.

SYND (Strategic Youth Network for Development). Environmental Sanitation Policy of Ghana, 2013. Accessed 5 July 2013. http://www.modernghana.com/news/447887/1/environmental-sanitation-policy-of-ghana.html.

Tamakloe, William. 2008. State of Ghana's Environment-Challenges of Compliance and Enforcement, 2008. Accessed 10 May 2013. http://www.inece.org/indicators/proceedings/04h_ghana.pdf.

Taringa, Nisbert. "How Environmental is African Traditional Religion?" *Exchange* ,35, (2) (2006): 191-214.

Taylor, Bron .R. "Ecology and Nature Religions." In *The Encyclopedia of Religion* edited by J. Lindsay New York: Macmillan, 2nd edn.,(2005): 261–66.

—. 2005. *The Encyclopedia of Religion and Nature.* London, New York: The Thomas Continuum.

Taylor-Powell, E and Renner, M. . *Analysing Qualitative Data*. 2003. Accessed April 6, 2013. http://learningstore.uwex.edu/assets/pdfs/G3658-12.PDF.

Technical Working Committee. "Recommendations for Controlling Illegal Chainsaw Milling in Ghana A two-day Technical Working Committee Meeting, within the Framework of the EU-funded project, under the theme: Developing Alternatives for Illegal Chainsaw Milling in Ghana and Guyana through Multi stakeholder Dialogue" (25-26 May 2009 in Accra, Ghana).Accessed July 5, 2013. http://www.illegal-logging.info/item_single.php?it_id=781&it=document.

Terborh, John. 1992. "Maintenance of Diversity in Tropical Forests" *Biotropica* 24 (1992):283-292.

Teye, Joseph. K. 'Network Management Of Water Resources in Ghana: A Framework for Integrating Traditional and Scientific Methods. Report of the National Conference on Culture and Water organised by the National Commission on Culture At Akosombo, 24- 26 February 2010.

The Assisi Declarations. Messages on Humanity and Nature from Buddhism, Christianity, Hinduism, Islam and Judaism, Basilica DI S. Francesco Assisi, Italy, WWF 25th Anniversary, 29 September 1986.

The World Bank .2012. "Poor Sanitation Costs Ghana GHC420 million each year", 2012. Accessed February 10, 2013. http://www.worldbank.org/en/news/press-release/2012/04/17/poor-sanitation-costs-ghana-ghc420-million-each-year

The Daily Graphic, 30 April 2013 edition.

The Ghana Statistical Service. The 2000 population and housing census.

The New Legon Observer "Environment, Conservation and Survival of the Species." 2 (8) (2008):3.

The Republic of Ghana, The 1992 Constitution.

Thorne, S. 2000. "Data analysis in qualitative research." *Evidence. Based Nursing*, 3(2000):68-70.

Timberlake, J., and Cunliffe, R. N. "Sites of interest for botanical conservation in the communal lands of the Zambezi valley, Zimbabwe. Consultant's Report for The Zambezi Society" The Zambezi Society, P.O. Box HG 744, Highlands, Harare, Zimbabwe, 1997.

Togarasei, L. "African Traditional Religion in the Study of The New Testament." In 2012. *African Traditions in the Study of Religion in Africa*. edited by A, Adogame,.E, Chitando and B. Bayete England, USA: Ashgate,(2012):205-218.

Tomalin, Emma. Religious Studies and Development: Literature Review (Religion and Development Research Programme Working Paper 6), 2007.

—. 2002. "The Limitations of Religious Environmentalism for India" *Worldviews: Global Religions, Culture and Ecology*. 6 (1) (2002):12-30.

—. "Tradition and Transformation: a Comparative Study of Religious Environmentalism in Britain and India" Ph.D. Thesis. The University of Lancaster, 2000.

Tropical Timber Report Since 1990-2006. *FC Cracks down on Illegal Teak Operators*. 11(11) (2006) :2.Accessed 10 January 2013]. www.itto.int/direct/topics/topics_pdf../topics_id=12780000&no=11.

Tuff our, K. "Sacred Groves: Traditional methods of Conservation" The *Daily Graphic*. Accra, June, 1991 Accessed January 4, 2010

Tucker, Mary. E. and Grim , John The Nature of the Environmental Crisis. In *Christianity and Ecology: Seeking the Well-Being of Earth and Humans*, edited by D. T. Hessel and R. R. Ruether . Harvard: Harvard University press, 2000.

—. Overview of World Religions and Ecology. 2009. Accessed June 2013 27, http://fore.research.yale.edu/religion/

Turaki, Y. *Christianity and African Gods: A Method in Theology*. Nairobi: IBS-Nig. Press, 1999.

Tylor. E.B. Primitive Culture: Research into the Development of Mythology, Philosophy, Religion, Art and Custom. London: Routledge, 1871.

UNCED (United Nations Conference on Environment and Development)-- *Agenda 21*. 3-14 June 1992, Rio de Janerio, Brazil.

UNESCO. 1996. Sacred sites- cultural integrity biological diversity programme proposal. November, Paris, 1996.

Underlying Causes of Deforestation:Africa Backgrpund Document.1999. World Rainforest-Movement. 1999. Accessed March 12, 2013. http://wrm.org.uy/deforestation/Africabackground.html.

USAID. "Biodiversity conservation: A guide for USAID staff and partners." 2005.

USHE, N.G.H. Between continuity and change: A comparative study of Tiv and Christian concept of death and burial rites PhD. Thesis., University of Jos, Nigeria, 2005.

Venema, L. B. Collaborating With Indigenous Farmers' Associations in Agricultural Development and Resource Management. In Local Resource Management in Africa, edited by J. P. M. Van Den Breemer, C. A. Drijver and L. B. Venema. Chichester: John Wiley and Sons Ltd.,1995.

Walker, P. "Decolonizing Conflict Resolution." *American Indian Quarterly*, 28 (3 & 4) (2004):.527-549.

Warren, Dennis M. "Comments on article by Arun Agrawal." *Indigenous Knowledge and Development Monitor,* 4(1) (1995):13.

—. *The Akan of Ghana: an Overview of the Ethnographic Literature*. Accra: Pointer Limited, 1986.

Warren, Dennis M., L.J. Slikkerveer and D. Brokensha (eds.) *The Cultural Dimension of Development: Indigenous Knowledge Systems*. London: Intermediate Technology Publications, 1995.

Westerlund, D. ""Insiders" And "Outsiders" in the Study of African Religions:Notes on some Problems of Theory and Method." In *African Traditional Religion in contemporary society*, edited by J.K. Olupona. St. Paul, Minneapolis: Paragon House, (1991):15-24.

Western Regional Directorate, Centre for National Culture. " Culture as a vehicle for wealth creation and socio-economic development." Report of the National Conference on Culture and Water organised by the National Commission on Culture at Akosombo, 24- 26 February 2010.

Wienecke, M.A. "Sustainability and the Role of Biodiversity: The Impact of Anthropogenic Activities." .*Journal for Sustainable Development in Africa*, 7 (2)(2005):70-95.

Williamson, Sidney. G. *Akan religion and the Christian faith*. Accra: Ghana Universities Press, 1965

White, Lynn. The Historical Roots of Our Ecological Crisis. S*cience* 155 (3767)(1967):1205-1207.

White, Leslie, A. with Dillingham, Beth. *The Content of Culture: Basic concepts in Anthropology*. Minneapolis: Burgess Publishing Company, 1972.

World Commission On Environment and Development (WCED) Brundtland Report and Sustainable development. Accessed December, 2009 .http://www.idrc.ca/en/ev-64525-201-1-DO_TOPIC.html, 1987.

Wolfensohn, J. D. Forward. In. *Faith in Conservation: New Approaches to Religions and the Environment*., edited by M. Palmer and V. Finlay, xi. Washington D.C.: World Bank,,2003.

Yanka, Kwesi. *Speaking for the Chief*. Indianapolis, 1995.

Yidana S M, Ophori D and Banoeng-Yakubo B. A Multivariate Statistical Analysis of Surface Water Chemistry – The Ankobra Basin, Ghana; *Journal of Environmental Management* 86 (2008):.80–87.

Zahan, D. Some reflections on African spirituality. In.*African Spirituality: Forms, Meanings and Experiences*, edited by J. K Olupona, 3-25. New York: The Crossroad Publishing Company, 2000.

Zinnbauer, Brian J., Kenneth I. Pargament, et al. "Religion and Spirituality: Unfuzzying the Fuzzy." *Journal for the Scientific Study of Religion* 36:4 (1997): 549–64.

INDEX